T0269618

CAMBRIDGE LIBRARY COLLECTION

Books of enduring scholarly value

Zoology

Until the nineteenth century, the investigation of natural phenomena, plants and animals was considered either the preserve of elite scholars or a pastime for the leisured upper classes. As increasing academic rigour and systematisation was brought to the study of 'natural history', its subdisciplines were adopted into university curricula, and learned societies (such as the London Zoological Society, founded in 1826) were established to support research in these areas. These developments are reflected in the books reissued in this series, which describe the anatomy and characteristics of animals ranging from invertebrates to polar bears, fish to birds, in habitats from Arctic North America to the tropical forests of Malaysia. By the middle of the nineteenth century, this work and developments in research on fossils had resulted in the formulation of the theory of evolution.

The Life and Letters of Gilbert White of Selborne

Published in 1901, this illustrated two-volume biography of the renowned English naturalist Gilbert White (1720–93) presents a thorough account of his life and achievements. Prepared by White's great-great-nephew Rashleigh Holt-White (1826–1920), it incorporates a selection of White's correspondence with family and friends, providing valuable insights into his beliefs and character. Included are letters sent by White's lifelong friend John Mulso (1721–91), who praised the naturalist's work, predicting it would 'immortalise' White and his Hampshire village. Still considered a classic text, *The Natural History and Antiquities of Selborne* (1789), featuring White's careful observations of local flora and fauna, is also reissued in the Cambridge Library Collection. In the present work, Holt-White sought to correct the 'erroneous statements' that had previously been made about his relative. Volume 1 covers White's life and achievements up to August 1776, including his studies at Oxford and his ordination as a priest.

Cambridge University Press has long been a pioneer in the reissuing of out-of-print titles from its own backlist, producing digital reprints of books that are still sought after by scholars and students but could not be reprinted economically using traditional technology. The Cambridge Library Collection extends this activity to a wider range of books which are still of importance to researchers and professionals, either for the source material they contain, or as landmarks in the history of their academic discipline.

Drawing from the world-renowned collections in the Cambridge University Library and other partner libraries, and guided by the advice of experts in each subject area, Cambridge University Press is using state-of-the-art scanning machines in its own Printing House to capture the content of each book selected for inclusion. The files are processed to give a consistently clear, crisp image, and the books finished to the high quality standard for which the Press is recognised around the world. The latest print-on-demand technology ensures that the books will remain available indefinitely, and that orders for single or multiple copies can quickly be supplied.

The Cambridge Library Collection brings back to life books of enduring scholarly value (including out-of-copyright works originally issued by other publishers) across a wide range of disciplines in the humanities and social sciences and in science and technology.

The Life and Letters of Gilbert White of Selborne

VOLUME 1

EDITED BY
RASHLEIGH HOLT-WHITE

CAMBRIDGE
UNIVERSITY PRESS

University Printing House, Cambridge, CB2 8BS, United Kingdom

Cambridge University Press is part of the University of Cambridge.
It furthers the University's mission by disseminating knowledge in the pursuit of education, learning and research at the highest international levels of excellence.

www.cambridge.org
Information on this title: www.cambridge.org/9781108076487

© in this compilation Cambridge University Press 2015

This edition first published 1901
This digitally printed version 2015

ISBN 978-1-108-07648-7 Paperback

This book reproduces the text of the original edition. The content and language reflect the beliefs, practices and terminology of their time, and have not been updated.

Cambridge University Press wishes to make clear that the book, unless originally published by Cambridge, is not being republished by, in association or collaboration with, or with the endorsement or approval of, the original publisher or its successors in title.

THE LIFE AND LETTERS OF

GILBERT WHITE OF SELBORNE

Gil. White Vicar.

THE LIFE AND LETTERS OF
GILBERT WHITE
OF SELBORNE

WRITTEN AND EDITED BY
HIS GREAT-GRAND-NEPHEW
RASHLEIGH HOLT-WHITE

WITH PEDIGREE, PORTRAITS, AND ILLUSTRATIONS

IN TWO VOLUMES

VOL. I.

LONDON
JOHN MURRAY, ALBEMARLE STREET
1901

TO

DAVID BINNING MONRO, M.A.

PROVOST OF ORIEL COLLEGE

FROM

A FORMER PUPIL

Floreat Oriel

PREFACE

I DO not propose to offer any apology for publishing a *Life of Gilbert White of Selborne*; indeed, I think that an apology is due to the Shade of a naturalist who holds so high a place in the opinion and regard —I might almost say the affection—of his countrymen, that no authentic account of his career has yet been given to the world.

Nevertheless, though a knowledge of natural history is not, of course, the only qualification necessary in a biographer of the Fellow of Oriel, the life of a naturalist should have been written by a naturalist; a title to which I have not the slightest claim: and I am not sure that I should ever have printed anything, had I not observed with regret that erroneous statements concerning the philosopher of Selborne were constantly occurring, in proportion to the interest taken in him which seems to be ever increasing.

That many of the circumstances of his life should be mis-stated; that, for instance, he should have been wrongly represented as remaining single on account of an unrequited attachment; that mistakes, sometimes of a rather ludicrous nature, should have been frequently made about his relatives, his habits, and the Selborne of his time by writers who have perforce drawn upon their imagination for their statements—all this might be passed by with a smile by those who know the truth. But it is hardly a laughing matter for one of his family to read in a recent guide-book to Oxford (by J. Wells, M.A., Methuen and Co., p. 98), that Gilbert White "held his Fellowship for fifty years, and a living into the bargain, though it was shrewdly suspected that his fortune exceeded what the Statutes allowed: he shocked even his contemporaries by his non-residence and pluralism"; a passage which I can only characterise as a gross libel that the writer's complete ignorance of the circumstances can barely extenuate.

It seems time, therefore, that the facts about Gilbert White's life and surroundings, so far as they can now be ascertained, should be placed on record; and I believe that the admirers of his graceful writings will be glad to have an authoritative account of the Selborne naturalist—a man whose character need fear no scrutiny.

If, in dealing with these facts, I have not in one instance been able to avoid the element of controversy, I must plead that

> "I speak not to disprove what Brutus spoke,
> But here I am to speak what I do know."

At the request of several correspondents I have added a pedigree of the family as far as the generation succeeding Gilbert White's; *i.e.* his nephews and nieces, who are often mentioned in his letters.

To my cousins who have kindly allowed me to copy letters and pictures; and especially to the Earl of Stamford, who has placed at my disposal the long, and for my purpose invaluable series of letters to the Naturalist from his friend and contemporary John Mulso, my best thanks are due; as well as to Mrs. Inge, a descendant of Archdeacon Churton, who has kindly lent me the letters to him from Gilbert White. The Norfolk and Norwich Naturalists' Society has been good enough to allow me to reprint from its 'Transactions' Gilbert White's letters to Robert Marsham. The Vicar of Selborne; the Rector of Fyfield; the present owner of "The Wakes"; Mr. Lazenby, of Basingstoke; and others, have kindly given me facilities.

Lastly, to Professor Newton my obligations are many and great. In addition to much valuable advice, he has been good enough to send me the natural history notes which appear with his initials.

R. H.-W.

New Year's Day, 1901.

CONTENTS

CHAPTER I.

CHAPTER II.

CHAPTER III.

CHAPTER IV.

CHAPTER V.

CHAPTER VI.

CHAPTER VII.

CHAPTER VIII.

CHAPTER IX.

CHAPTER X.

CHAPTER XI.

CHAPTER XII.

CHAPTER XIII.

LIST OF ILLUSTRATIONS

"How much," said he, "more happy is the State
In which ye, Father, here do dwell at ease,
Leading a Life so free and fortunate,
From all the Tempests of these worldly Seas,
Which toss the rest in dangerous Disease?
Where Wars, and Wrecks, and wicked Enmity
Do them afflict, which no man can appease;
That certes I your Happiness envy,
And wish my Lot were plac'd in such Felicity."

"Surely, my Son (then answer'd he again)
If happy, then it is in this Intent,
That having small, yet do I not complain
Of Want, ne wish for more to it augment,
But do my self, with that I have, content;
So taught of Nature, which doth little need
Of foreign Helps to Life's due Nourishment.

. . . . .

"To them that list, the World's gay Shows I leave,
And to great ones such Follies do forgive,
Which oft thro' Pride do their own Peril weave,
And thro' Ambition down themselves do drive
To sad Decay, that might contented live.
Me no such Cares nor combrous Thoughts offend,
Ne once my Mind's unmoved Quiet grieve;
But all the Night in silver Sleep I spend,
And all the Day, to what I list, I do attend."

The Faery Queen.

"Your work, upon the whole, will immortalize your place of abode as well as yourself."

JOHN MULSO; 16*th July*, 1776.

THE LIFE AND LETTERS OF

GILBERT WHITE

OF SELBORNE

CHAPTER I.

THOMAS, brother of Gilbert White of Selborne, who, after his retirement from the cares of business in 1777, amused himself by writing letters on various topics (chiefly under the signature "T. H. W.") to the "Gentleman's Magazine" and other periodicals, wrote, in 1786, to the one just named regarding the early history of the Whites of Hampshire as follows:—

MR. URBAN,—As I value myself on being a descendant of the *Jutæ* or *Viti*, I am obliged to take notice of the reflection that Governor Pownall casts on my ancestors by calling them *pirates* ("Archæologia," vol. vii. p. 269). We Guti or Viti were permitted by King Ina to settle in this kingdom on an equal footing with the rest of his subjects, because we are descended, as we can prove by record, "*de nobili sanguine Anglorum,*" *from the noble blood of the Angles,*

(Leges Edovardi) . . . give me leave to say that few families have so antient or equitable a title to their possessions in this island. They who came in with the ravaging Danes or with the Norman invader certainly have not; and if the early Saxons had committed any injustice in their first establishment, it was before we migrated hither, for we had the rare felicity to settle peaceably and to be admitted to all the privileges of fellow-citizens with general content. Where is the man, unless he can prove his descent from the *Armorici*, who can make it appear that his ancestors gained an establishment in this country on terms so respectable?*

RICARDUS VITUS BASINSTOCKIUS.†

It is true that the earlier Jutish colonies were founded by emigrants, as contrasted with those of the Angles, who came in a body as conquerors. The name of White is a common one, especially in Hampshire, Berkshire, and Oxfordshire, and this points to the probable conclusion that it is in its modern form a corruption of a tribal name. But however this may be, leaving early history, it appears from the Harleian MSS. that there were in the late Middle Ages four considerable families of that name established in Hampshire, who seem to have been related both by descent and intermarriage.

1. White of Swanborne, or South Warnborne.
2. White of Basingstoke.
3. White of Farnham.
4. White of Aldershot.

* "Gentleman's Magazine," 1786, vol. lvi. p. 17.

† The reason of the adoption of this name appears *infra* p. 6.

Anthony a Wood's MS., now in the Bodleian Library at Oxford, after giving an account of Gilbert White's great-grandfather, Sir Sampson White, Knight, and his family, has a note (in Wood's handwriting, but added apparently at a later date) to this effect:—

"These Whites are descended from the Whites of South Warnborough, in Hampshire. The original of the White of Oxfordshire was steward or Baylier of Einsham Abbey."*

This note rather conflicts with the account given previously in this MS., and it is to be regretted that no authority for the later statement is given.

The pedigree of the Whites of South Warnborough given in the Harleian MSS.† commences with one Robert White, born at Yatley, Hants, who was "Mayor of the Staple at Calais, lived at Sandwich and after at Farnham, and purchased the manor of Southwarnborne, co. Hants, of Sir Fulke Pembridge." He died at Farnham in 1461.

His son John, who was knighted, married a daughter of Robert, Lord Hungerford; and their son Robert, who married a daughter of Sir Thomas Inglefield, the elder, Knight, was succeeded by his eldest son Sir Thomas White, Knight, Master of the Requests to Queen Mary and Sheriff of London and Middlesex, 1 Edward VI. He married Agnes, daughter of Robert White of Farnham and sister of John White, Bishop of Lincoln, and subsequently

* *Vide* note to Wood's MS., f. iv. p. 155.

† Nos. 1,544, 1,183, 1,139, etc.

of Winchester in Queen Mary's reign, and of Sir John White, Lord Mayor of London in 1563. Sir Thomas's eldest son Henry possessed the manor of South Warnborough; but with him the family became extinct at South Warnborough in the male line, as his only children were three daughters, coheiresses, who married respectively Thomas Scudamore, Walter Gifford of Chillington, Stafford, and Henry Ferrers of Badisley Clinton, Warwickshire. John White, a grandson of Sir Thomas, was declared chief heir male.*

The manor-house of South Warnborough, though now much reduced in size and altered, existed in its original state until about 1830, and was a fine red-brick Tudor mansion. The chancel of South Warnborough Church is filled with brasses and monuments, some of them fine and interesting, of the White family, which intermarried with the Paulet, Gaynsford, Tichborne, and other old Hampshire families, and evidently occupied a position of some importance in the county.

In the church windows are several coats-of-arms, in stained glass, of the family and its alliances, which were formerly in the manor-house—probably

* MS. Harl., 1,544. He was living in 1593. In "Nature Notes," vol. iv. p. 106, this John White is represented as identical with John White of Coggs, near Witney, the father of Sir Sampson White; but this is an error, since Thomas (not Richard, as stated in "Nature Notes") White, the father of the former, in his will dated September 1st, 1558, describes himself as "of Downton in Wilts gentleman," and speaks of his son John as his only child, whereas John White of Coggs mentions a brother Henry in his will.

in the large hall, now pulled down, which is said to have witnessed the knighting of one of its owners by Queen Elizabeth.

The arms of this branch of the family, the grant of which is duly recorded, are (with a slight difference, which evidently arose from a mistake on the part of the Oxfordshire family) the same as those which appear on the handsome monuments in S. Mary's Church, Oxford, to Sir Sampson White and his sons, and to his brother Henry in the church of Coggs, near Witney; viz. argent,* a chevron gules between three parrots or popinjays vert, collared or, a bordure azure besantée.

The Whites of Basingstoke and Farnham were descended respectively from Jenkin † and John, the two sons of Thomas White of Pernix, or Purvile, in Hants.

Of the former family the most celebrated member was Richard White, the son of Henry White of Basingstoke. He was educated at Winchester and New Colleges, of which latter he was admitted Fellow in 1557. Leaving England on account of his religious opinions, he went to Louvain, and afterwards to Padua, where he became D.C.L. Subsequently he was created by the Emperor a "Comes Palatinus," and by the Pope "Magnificus Rector" of

* In the Oxfordshire and earlier Selborne monuments the shield is *or*.

† A. a Wood states that Jenkin White "had almost half the town of Basingstoke in his possession."

Douay, at which place he died a canon of S. Peter's Church in 1611.

He wrote under the name of *Ricardus Vitus Basinstockius,* his principal work being a History of the British Isles, "ab origine mundi." He also wrote some orations; two of which, spoken at Louvain, were published by Christopher Johnson, Headmaster of Winchester College, about 1564, and commanded by him to be read publicly in the school by the scholars; also a geometrical treatise, entitled "Hemisphaerium Dissectum,"* and other works. With Richard White the Basingstoke family seems to have become extinct in the male line.

The Farnham branch of the family produced a man of some eminence in John White (son of Robert White of Farnham), who was born at that place in 1511. According to Anthony a Wood,† he was "educated in grammar learning at Wykeham's School, near Winchester, Fellow of New College in 1527, M.A. in 1533 (and about that time master of the said school), Warden in 1541."

In Queen Mary's reign he was Bishop of Lincoln, and subsequently of Winchester. He "helped burn Ridley, preached a black sermon on Queen Mary's death of Elizabeth, 'melior est canis vivus leone

* In the Imprimatur appended to this mathematical dissertation (which is gravely stated to contain "nihil contra fidem aut bonos mores") the author is described as "Richardus Albius nobilis Anglus praeillustri et antiquissima familia apud Anglos natus." In this book his arms are represented as those of the South Warnborough family.

† 'Athenæ Oxonienses' (Ed. Bliss), vol. ii. p. 118.

mortuo,' and threatened to excommunicate the Queen, who committed him to the Tower in April, 1559, and deprived him of his bishopric." He was shortly afterwards, however, permitted to retire to his sister's, Lady White's, house at South Warnborough, where he died on the 11th January following.*

Another son of Robert White of Farnham, Sir John White, Lord Mayor of London in 1563, founded a branch of the family at Aldershot. His grandson Robert White, however, being the only surviving male member of his family and dying without issue, the family became then extinct in the male line. Robert White's two sisters married two brothers, Sir Richard Tichborne of Tichborne, Bart., and Sir Walter Tichborne of Aldershot, Knight.

It has been stated that according to Anthony a Wood the Whites of Oxfordshire, who were undoubtedly the immediate ancestors of the Whites of Selborne, were descended from a member of the South Warnborough branch of the family. Following, however, Wood's earlier account we find that he traced his contemporary, Sir Sampson White, Mayor of Oxford, from a family settled at "Cogges

* It may perhaps be of some interest to note that when, a few years ago, eight apostle spoons which had belonged to Bishop John White were sent to Christie's by descendants, in the female line, of Dame Agnes White, they fetched no less than 265 guineas. It was stated that had the other three spoons and the "king" spoon been also sold, the lot would probably have fetched 1,000 guineas.

by Witney, where the name hath lived several generations."

Gilbert White, the naturalist, a great-grandson of Sir Sampson, writing for the information of a nephew, Samuel Barker, of Lyndon Hall, Rutland, gives the following account of his ancestors:—

"The family of the Whites (our family) were in possession of an estate called Swan Hall, in the tything of Haley, parish of Witney, Oxon., as long ago as the reign of Queen Elizabeth; the mansion of Swan Hall is still standing. In the map of Oxfordshire bound up with Plot's history of that county* are the arms of the Whites of Swan Hall, the same that we bear. We are lineally descended from Sir Sampson White, my great-grandfather, being the fourth son (born 1607), whose father was possessed of Swan Hall. This estate, by a female line, went into the family of the Ashworths, who sold it."

As already stated the name of White is by no means uncommon in Oxfordshire. Among the names of the gentry of Oxfordshire returned by the Commissioners in the twelfth year of King Henry VI. occurs Johannes White.† One John le Whyte's name occurs in Dugdale's 'Monasticon Anglicanum' as a witness to a deed relating to Einsham Abbey.

It is certain that a family of that name was settled at Coggs in the reign of Henry VIII. In the Sub-

* 'The Natural History of Oxfordshire,' by Robert Plot, LL.D., late Keeper of the Ashmolean Museum and Professor of Chemistry in the University of Oxford, 1677.

† *Vide* FULLER's *History of the Worthies of England.* Ed. Nicholls. Vol. ii. p. 235.

sidy Rolls for the hundred of Wotton, co. Oxon., of November 29th, 16 Hen. VIII. (1524), in the list of names for the village of Coggs occur—

	s. d.
"Willm Whyght . . .	vi. viii.
Alicia Whyght, vidua. . .	iiii.
Thomas Whyght . . .	iiii."

These names appear, from an examination of Alicia Whyght's will, made in 1531 and proved in 1553, to be those of a mother and her two sons, who were grown up, and householders in 1524. This family seems to have been amongst the most substantial people in Coggs, since there is only one person assessed at a higher rate.

The names of Thomas Whyte and Richard Whyte of Coggs are conjectured from an examination of their wills to be grandfather and father respectively of John White of Coggs, who made his will in 1628, and who had four sons and two daughters.

His youngest son Sampson, who was born in 1607, went to Oxford and became a mercer there. He prospered in business, and became, according to Wood,

"Baillive of the city of Oxon. 1642, turned out of the Councell House for his loyalty, restored upon His Majesties coming to the Crowne 1660, chose mayor y^t yeare, served as Butler of the Beer-celler at his Coronation, and then had the dignity of Knighthood confer'd upon him. In 1665 he was chose Mayor againe, being ye yeare w^n the great plague raged in London."

An interesting "account of the claim of the city of Oxford made at the coronation of King Charles the Second" is preserved in the city book.

It appears that "Certeyne Lords Com^{rs}" were appointed

"for the recieving the peticōns & claimes of such as were to doe service on the said day. Hereupon Sampson White, Esq., Mayor of the City, by the advice of his Brethren and the whole Councell Chamber, took his journey to London to present from this city a peticōn & claime to the Com^{rs} that the Mayor of Oxford and six of the Cittizens might be admitted to serve the King's Ma^{tye} in his Buttelary on the day of his Royall Coronacōn together with and in such manner as the Lord mayor and the cittizens of London doe."

This petition, drawn up in Norman-French, was allowed by the commissioners. Whereupon the Mayor, "with 6 Cittizens appointed by the Councell chamber" and "5 others of the councell chamber who offered to habit themselves att their owne charge," with the mace-bearer and sergeants, "rode up to Westminster for to doe the service aforesaid." Upon their application on the evening before the coronation, the Duke of Ormonde, "steward of his Ma^{tyes} household," formally put "the Mayor and cittizens of Oxford into possession of the King's Butelarryes." When the King returned from Westminster Abbey to the feast in Westminster Hall,

"Mr. Maior accompanied with Mr. Ernley, Gent of the Buttery, went up in his scarlett gowne from the Buttery at the lower end of Westm^{r} Hall to the cubbard erected

for the Butler neer to the King's table carrying two golden Bowles covered in his Hands wth wch his Majesty was to be served . . . . at which cubbard Mr. Mayor of Oxon: alone waited in his scarlett gown, executing the office of Buttler while his Matie was at dinner only, when his Matie had neer dyned the Lord Mayor of London presented his Matie with a bowle of wine & claimed the bowle & had the same allowed by his Matie.

"And it is to be noted that the Friday Morneing Mr. Mayor and the Cittizens attending on his Matie to kiss his Royall Hand his Matie was graciously pleased to conferre the hon^{r} of K^{nt}hood upon Mr. Mayor."

It is mentioned that on their application the Mayor and his followers were permitted by the Lord Mayor of London "to go down to Westminster with his Lordshippe in his Barge" on the Coronation day, after a "very free Treatment given them by his Lsppe." They also dined with the Lord Mayor on the day after the coronation, "where they had a very noble & great feast & were entertayned by his Lsppe wth soe much and soe great respect that it deserves to have a lasting rememberance for wch purpose it is here entered."

The "bowle," a large silver-gilt cup, which was kept by Sir Sampson White, was bequeathed by him to the Corporation of Oxford.

A family tradition, which has descended to the writer, records that the worthy Mayor, on hearing Lord Rochester at Court remark on the recent frequency of highway robberies, sagely observed,

"Perchance then, my lord, there be knaves abroad!" Upon which his lordship, looking at him, replied, "Ay, and fools too!"

Sir Sampson White's eldest brother John is described as "of Northley, near Witney." The next brother, Richard, became Vicar of Basingstoke; and the next, Henry White of Coggs, died *cælebs* in 1677, and was buried at Coggs. In this church there is a monument to him with a Latin inscription under his coat-of-arms, in which he is stated to be "ex familiâ vicum hunc (ultra certam temporis prescriptionem) habitante." The inscription records his travels through various parts of England, his readiness to impart information thus acquired, his charity, etc., and states that the memorial was erected by his nephews Henry, Francis, and John—three of Sir Sampson's sons, one of whom was his heir under his will.

Possibly this member of the family may have owned Swan Hall; but it seems almost certain that Gilbert White was mistaken in asserting that Sir Sampson White's father, John White, owned it, since there is no mention of this estate in his (John White's) will. It seems more probable that his son Sir Sampson, who became the wealthy man of the family, bought this property in his native village of Coggs, with a view to one of his sons living there.

Sir Sampson White, who married Mary, daughter of Richard Soper, Arm., of East Oakley, in Hampshire, died "early on Sunday morning 28 Sep[r] 1684 at his house in S. Marie's parish, opposite to Universitie College, aged 78, and was buried on the 5th Oct. in S. Marie's Church, viz. between the dore leading thence into Adam Brome's chappell and the buttress of the steeple, on the north side of this church."

A long Latin inscription under his coat-of-arms records his dignities and virtues, and states that he was "ab omnibus desideratus." *

He left a family of five sons, three of whom became graduates of the University, and one daughter. John White, born 1636, his eldest son, of Magdalen and afterwards of University College, M.A., J.P.; Richard White, born 1647; Henry White, born 1648, who succeeded his father in his business and also became Mayor of Oxford. He built a large house at Chalgrove, near Warborough, Oxfordshire. Gilbert, born 1650, M.A., Fellow of Magdalen College; and Francis, born 1652, M.A., B.D., Fellow of Balliol College.

Some mention is made by Wood of these sons of Sir Sampson. John White is spoken of as a justice of the peace and living in the parish of Witney, † "but the said John White, showing himself too malapert when the popish plot broke out, was left out of the commission [of the peace] 1679."

* Sir Sampson White's hatchment, together with that of his son Gilbert, was preserved in Selborne Church, where they were seen in 1844 by a member of the author's family, They have since disappeared.

† At Swan Hall apparently. *Vide* map in PLOT's 'Natural History of Oxfordshire,' in which the arms of "White, Esq.," are referred by number to Haley, in Witney parish.

"The next brother to John is Henry White," Mayor of Oxford in 1691–2, who seems to have been called "Squire White." After him, "The next son is Gilbert White, Fellow of Mag$^{d.}$ College, afterwards Rector of Selbourn, in Hampshire, by y^{e} gift of his College." He held his Fellowship from 1672–81, being presented to the living of Brandeston, Norfolk, 21st June, 1680. In February, $168\frac{0}{1}$, he was presented to Selborne, Hants, the Register of Baptisms of which church contains the following entry in his handwriting:—

"Gilbert White took possession of y^{e} Church June y^{e} 23^{d} 1681."

His grandson, in his "Antiquities of Selborne,"* notes that at this time the living of Selborne was a very small vicarage, and after giving a list of vicars, says:—

"June 1681. This living was now in such low estimation in Magdalen College, that it descended to a junior fellow *Gilbert White*, M.A., who was instituted to it in the thirty-first year of his age."

In what humble circumstances the rural clergy were sometimes accustomed to live at this date is apparent from the following:—

"At his first coming he ceiled the chancel, and also floored and wainscoted the parlour and hall, which before were paved with stone, and had naked walls; he enlarged the kitchen and brewhouse, and dug a cellar and well: he also

* Letter VI.

built a large new barn in the lower yard, removed the hovels in the front court, which he laid out in walks and borders; and entirely planned the back garden, before a rude field with a stone-pit in the midst of it."

The vicar married Rebecca Luckin, the daughter of a yeoman of Nore Hill, in the neighbouring parish of Newton Valence. She outlived her husband some years, dying in 1755 in her 91st year. He died in 1727, and his virtues and charities to the parish were commemorated by a tablet in the chancel of the church.*

Of their children, who lived to maturity, their son John was born in 1688 and baptised at Burstow, in Surrey; Mary, born in 1689, married the Rev. Baptist Isaac, Rector of Whitwell and Ashwell, Rutland; Rebecca, born in 1694, married Henry Snooke of Ringmer, Sussex, Esquire; Dorothea, born in 1696, married the Rev. W. H. Cane, her father's successor in the vicarage of Selborne; and Elizabeth, born in 1698, married the Rev. Charles White, Rector of Bradley and Swarraton, who was a cousin.

John White, of the Middle Temple, Barrister-at-law and J.P. for Hants, the vicar's only son who lived to maturity, married on September 29th, 1719,

* It is here recorded that "he consecrated the tithe of his fortune to pious uses," viz. £950, which he specifically bequeathed "to charitable uses" by his will. *Inter alia* he left £40 for repairs to the church, £200 for repairs of the parish highways, and £100 for purchase of land, "the yearly rents whereof shall be employed in teaching the poor children of Selborne parish to read and write and say their prayers and Catechism, and to sew and knit."

at Rogate, Sussex, Anne, daughter and heiress of Mrs. Holt, of East Harting, Sussex, the widow of the Rev. Thomas Holt, Rector of Streatham, Surrey. Mrs. Holt was daughter and heiress of Benjamin Hyde (or Hide), of Chichester, Esquire, a Russia merchant; whose father, Benjamin Hyde, in return for some kindness shown to King Charles I. when a prisoner in Carisbrook Castle, received from the King his ribbon of the order of the Garter.*

Benjamin Hyde, jun., Mrs. Holt's father, had married Anne Ford, daughter of the second son of Sir William Ford of Winchester,† who brought him property at East Harting, which descended to Mrs. Holt.

Upon his marriage with Miss Holt, John White retired from London and the Bar, where his practice had not become considerable.

It seems probable that he was at first without a fixed residence, for in the summer of the year

* The ribbon, which is still in the possession of a member of the White family, is enclosed in a paper inscribed, "This is the original George ribbon of King Charles I., and given to Mr. Benjamin Hide, when the King was a prisoner in the Isle of Wight, and kept in memory of that good King." Mr. Hyde was certainly not "governor of Carisbrook Castle" as stated by Mr. Bell (vol. i. p. xxii.); but probably, as a shipowner, was a party to one of the abortive attempts to escape to France planned for the King.

† The Ford family was of some note in Hampshire and Sussex. Anne Ford's uncle, Sir Edward Ford of Up Park, Harting, had a daughter Catherine, who, marrying Ralph Grey, Arm., became the mother of Lord Grey of Werke, created Viscount Glendale and Earl of Tankerville by William III. A Lord Tankerville purchased property at Harting, inherited through Anne Ford, of John White of Selborne, and his wife.

Walker & Cockerell, ph. sc.

Anne Holt

following that of his marriage he and his wife were staying with his father at Selborne Vicarage; and in this house, on July 18th, 1720, was born his eldest son;† who, though he lived the quiet life of a scholar and naturalist, was destined to make the name of his secluded and obscure native village a household word wherever the English tongue is spoken; to link it with his own, and to achieve posthumous fame as Gilbert White of Selborne.

† *Vide* Gilbert White's little poem, "The Invitation to Selborne," l. 61 *et seq.*—

> "Nor be the Parsonage by the muse forgot,
> The partial bard admires his native spot."

He was christened in Selborne Church on August 1st, 1720. It has been assumed, but without any real authority, that he was a godson of the vicar, who mentions one grandson only, John, as "my godson" in his will.

CHAPTER II.

After quitting his father's vicarage at Selborne, John White and his wife first settled at Compton, about four miles from Guildford, in Surrey, and there remained till 1727, when they removed for a short time to East Harting, probably to Mrs. Holt's house there, where their son Francis was born in March, $172\frac{8}{9}$. Shortly afterwards they removed to Selborne, where they took up their residence at the house then known as "late Wake's" with John White's mother, who had been left a widow by the death of the Vicar of Selborne on February 13th, $172\frac{7}{8}$, and had consequently removed to that house from the vicarage. Here their two youngest children, Anne and Henry, were born. Mrs. John White died in 1739 at the comparatively early age of 46, after an attack of measles. Her husband did not marry again.

Gilbert White's relations with his brothers and sisters were so constant and pleasant, and he was so much indebted to his brothers Thomas, Benjamin, John, and Henry for sympathy and assistance in his studies in natural history; that, at the risk of some repetition, it will be well to give here a brief account of them.

His eldest brother *Thomas*, born at Compton, in Surrey, on October 13th, 1724, was educated at Bishop's Waltham, Hants, by Mr. Hampton (father of the translator of "Polybius"), and he is described by his nephew, John White,* as "an elegant classic." He went into a wholesale merchant's business in Thames Street, London, as a young man, becoming a partner of Mr. William Yalden, a brother of the Rev. Richard Yalden of Newton Valence, the adjoining parish to Selborne, who was long the neighbour and friend of Gilbert White. In 1758 he married Mr. W. Yalden's childless widow. In 1759 a daughter, Mary, was born, and in 1763 his wife died, shortly after giving birth to twin sons. In 1776 he came into full possession of considerable estates in Essex through the bequest of Mr. Thomas Holt, a half-brother of the Rev. Mr. Thomas Holt, of Streatham, and therefore his maternal great-uncle in the half-blood, whose name of Holt he then prefixed to his own surname, and, retiring from business soon afterwards, settled at South Lambeth, where he lived until his death. In January, 1777, he was elected a Fellow of the Royal Society. He was also a "citizen and apothecary." During the latter

* In a pedigree of the family compiled by him in 1826, now in the author's possession.

years of his life he was a constant contributor upon the most varied subjects to the "Gentleman's Magazine," as already stated, chiefly under the signature of "T. H. W.," a series of articles upon "The Trees of Great Britain" being said to be "particularly deserving of notice for the extensive information, good taste, and variety of reading which they display." * In conjunction with his brother-in-law, Thomas Barker, he communicated several papers to the Royal Society. A good botanist, an Anglo-Saxon scholar, and interested in antiquities, he assisted his brother in the preparation of his book; indeed, it was largely at his instance and owing to his solicitations that "The Natural History and Antiquities of Selborne" was prepared for publication. He died in February, 1797.

Benjamin, the next brother, born at Compton on September 15th, 1725, was also educated at Bishop's Waltham. Presumably through the circumstance of the marriage of his sister Anne to Mr. Barker, who was a grandson on his mother's side of the celebrated William Whiston, he entered into partnership with a Mr. Whiston, who was a relation of the last-named, as a publisher and bookseller at the sign of the "Horace's Head," No. 51, Fleet Street. He became the chief publisher of his time of works relating

* The late Thomas Rivers, of Sawbridgeworth, the pomologist, was struck with the knowledge displayed in these essays on trees, which were shown to him in 1855 by their author's grandson, the late Algernon Holt-White.

to natural history, and prospered in business, retiring to the then rural village of South Lambeth about the same time as his brother Thomas, near whom he lived there, and ultimately to a house which he hired of Lord Stawell, Marelands, in the parish of Bentley, Hants, a few miles from Selborne. Here he died in March, 1794, and was buried at Selborne near his brother Gilbert. He married first, in 1753, Anne Yalden, a sister of the Rev. R. Yalden, Vicar of Newton Valence; and secondly, in 1786, Mary, the widow of the Rev. R. Yalden. By his first wife he had a numerous family, and was succeeded in business by his sons Benjamin and John, the latter of whom records that his father was "citizen and Merchant Taylor, drank to as Sheriff in Alderman Sainsbury's mayoralty, 1787." His son Edmund became Vicar of Newton Valence in 1784, after the death of his uncle, Richard Yalden.

Rebecca, the eldest daughter of John White who grew up, was born at Compton on October 24th, 1726. In 1761 she married Mr. Henry Woods, a merchant in London, son of John Woods, Esq., of Chilgrove, near Chichester, Sussex. The house of his relations (by marriage) at Chilgrove is frequently mentioned by Gilbert White, who used often to stay there on his annual journey along the Downs to his aunt at Ringmer. Mrs. Woods died in 1771.

John, the next brother, born at Compton on September 20th, 1727, was probably little inferior even to his eldest brother Gilbert as a naturalist. He was elected to a (Surrey) scholarship at Corpus Christi College, Oxford, where he was admitted on March 12th, 174$\frac{5}{6}$, and took his B.A. degree in 1749. He was ordained in 1753 and became a curate at Barnet and in London. In 1754 he married, and in 1756 proceeded to Gibraltar as chaplain to the garrison there. While abroad he maintained a constant correspondence with his brother Gilbert, and towards the end of his sojourn on the Rock wrote to and received some letters from Linnæus upon natural history subjects, concerning which he accumulated much material. Returning to England in 1772, he became, on the presentation of the Archbishop of Canterbury, Vicar of Blackburn, and there prepared his natural history notes and caused illustrations to be made with a view to their publication, which, however, never took place.* He died, after a long illness, at Blackburn, in November, 1780. After his death his widow came to reside at Selborne with her brother-in-law, whom she outlived.

Of another brother, *Francis*, there is but little to say. He was born at East Harting, Sussex, on March 3rd, 172$\frac{8}{9}$. "Brother Franky," in London, is mentioned in a letter of Gilbert White's written in 1746. In a pedigree of the family drawn up by James White (son of Benjamin) and corrected by

* The probable fate of the work on the Fauna of Gibraltar is discussed *infra*, vol. ii. p. 66.

his uncle Gilbert, it is stated that Francis White married and had one daughter, who died an infant, and that he died in 1750 and was buried at Islington.

Anne, the next daughter of John White, was born at Selborne on April 2nd, 1731. She married Thomas Barker, of Lyndon Hall, Rutland, the representative of a collateral branch of an ancient family, of which Sir Abel Barker, Bart., of Hambleton, in the same county, was the head. Mrs. Barker had one son and three daughters. Frequent visits and much correspondence were exchanged between the Barker family and Gilbert White, Mr. Barker himself being a man of some literary and scientific knowledge, and his only son, Samuel, an apt disciple of his uncle in natural history researches. Mr. Barker died in 1809, and his wife in 1807.

The youngest son, *Henry*, was born at Selborne on June 25th, 1733. He was educated by Dr. Woodeson at Kingston-on-Thames, and under his eldest brother's auspices proceeded to Oriel College, Oxford, where he took his M.A. degree. He took Orders and obtained the (Chancellor's) living of Tidworth, Wilts. In 1762 he became Rector of Fyfield, and also subsequently Vicar of Uphaven, Wilts. His brother Gilbert was a frequent visitor at Fyfield, from which place some of his letters to Pennant are dated. Henry White married a Miss Cooper, of Oxford, by whom he had a numerous family. Several of his nephews and a few strangers were received by him at Fyfield to be educated with his own children. He died at Fyfield in December, 1788.

The village and neighbourhood of Selborne have been so often described, and are so well known to Gilbert White's disciples and admirers, that it is not necessary to very particularly dwell upon them here. Nevertheless, since its position of great seclusion, and certain of its natural features had undoubtedly much to do with the life and habits of its ardent admirer and lifelong resident, Gilbert White, the subject cannot be entirely disregarded.

The village lies, then, almost equidistant about five miles, from the Southampton and Portsmouth coach roads respectively; and in the naturalist's time, and for many years afterwards, it could only be reached from these roads through one of three lanes, as regards which the traveller would probably have been puzzled to say which he considered to be the worst. Turning from the main road to Southampton at the little town of Alton, he might go along a narrow, winding lane past West Worldham to Harteley, whence he would travel (upon the native rock!) for about two miles between steep and almost perpendicular banks, so near to each other that carriages could not pass one another except at particular points. This is one of the "two rocky hollow lanes" described at the beginning of the fifth letter to Pennant Bad as the author of these letters knew this lane to be, he used it sometimes, as appears from some of his private letters, even in winter. In truth, the traveller in his day had but little choice of a better route. He might, indeed, turn off the main road before reaching Alton, at Farnham, and take that towards Petersfield; then, skirting Wolmer Forest, he would find himself approaching Selborne through the hamlet of Oakhanger, whence, leaving the sandy forest ground, he could reach Selborne through the second "rocky hollow lane." This lane had, indeed, as Gilbert White records,* benefited from the bequest of a sum of £200 by his grandfather, which "was carefully and judiciously laid out in the summer of the year 1730 by his son John White, who made a solid and firm causey from *Rood-green*, all down *Honey-lane*, to a farm called *Oakwoods*, where the sandy soil begins."

This, however, was rather a roundabout way to Selborne, and probably the remaining lane, which seems to have been more generally used, was the most practicable route to the village. Following the Southampton road past Alton to East Tisted, the traveller turned from it at that village, at the Horse and Jockey Inn, and entered Selborne through a rough country road, which, though not so long as the Harteley lane, had nevertheless the disadvantage of a steep and stony hill to be descended.

There was, and indeed is, a fourth road, or rather

* *Vide* "The Antiquities of Selborne," Letter VI.

lane, of a rough description to Selborne, over the hills from Greatham on the Farnham and Petersfield road, but it would not have been used by visitors travelling from London.

The seclusion of Selborne residents, consequent on the wretched state of its roads, may well be gathered from the following significant entry made by Gilbert White in his journal, or *Garden Kalendar* as he termed it, under date March 15th, 1756:—

"Brought a four-wheel'd postchaise to ye door at that early time of year."

What a revelation of county life in winter at this time, especially for the ladies of the family, this entry affords!

It may be remarked that when he had chosen his road and was prepared to encounter its difficulties, the traveller's troubles were by no means at an end. Gilbert White's college companion, John Mulso, though he had frequently visited his friend at Selborne, regularly wrote for a guide to meet him "at the cross-roads," remarking that Selborne was as difficult of access as Rosamund's bower, and comparing his Selborne friends to "toto divisos orbe Britannos."

When the village was, however, at length reached through one of the difficult and winding lanes, a person who had any eyes for natural beauty must have been hard to please, if he were not charmed with its appearance. A small, very clear stream of water emerging from the chalk has hollowed out a little wooded valley, and flows roughly from west to east, past what was in the eighteenth century an old timber-framed vicarage, and its garden with

> ". . . its scapes grotesque and wild,
> High on a mound th' exalted gardens stand,
> Beneath, deep vallies scoop'd by Nature's hand," *

past the old church with its low, square tower, and through the meadows—called the long and short lythes—and so past the site of the ruined priory to Farnham and the River Wey. On the south side of this little valley runs the village street of thatched, timber-framed cottages, dominated, a little further south again, by an abrupt hill, perhaps 300 feet in height, clothed with a beech wood, or "Hanger." From the top of this eminence a charmingly timbered, park-like down stretches southwards, and affords extensive and very beautiful views in all directions over a hilly country—the long line of the downs above Portsmouth to the south, the hills about Midhurst and Haslemere on the east, Wolmer Forest, and Hindhead to the north-east, and the rolling and wooded upland country from Alton towards Alresford on the west, being prominent objects of the view. A land of woods and streams, and therefore an especially good country for birds—a character which it retains to this day.

Always happy in his quotations, the historian

* *Vide* "The Invitation to Selborne," by Gilbert White.

of Selborne well described his neighbourhood in the lines in which Ulysses pictures his native Ithaca, which he inserted on the title-page of his "Natural History"—

> "Τρηχεῖ', ἀλλ' ἀγαθὴ κουροτρόφος, ὄυτι ἔγωγε
> Ἧς γαίης δύναμαι γλυκερώτερον ἄλλο ἰδέσθαι." *

Fronting to, and opening upon the little village street, almost opposite the "Plestor" and church, stood a modest house somewhat irregularly built of stone edged with red bricks. This house, apparently erected about the end of the seventeenth century, is now known by the name of "Wake's," or "The Wakes." It had at the back a pleasant parlour on the first floor, from whose low window-seats could be seen a garden opening into several little fields; which, dotted with trees singly and in small clumps, stretched up to the dark and towering beechen Hanger; a charming little park-like territory, the home and playground of the various birds whose history so largely engaged the attention of the philosopher of Selborne.

Here he spent his boyhood after his parents settled in Selborne in his tenth or eleventh year; and this house, in spite of occasional absences owing to the exigencies of school and university and two or three short curacies (to say nothing of his numerous journeys to different parts of England), he never ceased to regard as his much-loved home during his whole life.

* **"A rough but good nursing-mother, nor can I see a sweeter land than this."**

CHAPTER III.

Of Gilbert White's early years there is indeed little on record. For people who are interested in such trifles it may be recorded that he was "bred up by hand," as he mentions in a letter to his niece Mary, written in 1784. Some part of his boyhood was spent at Farnham, a fact which he recorded in his "Naturalist's Journal" when staying, in the last year of his life, with his brother Benjamin, who had retired to Mareland, in the parish of Bentley, near that town: "The sweet peal of bells at Farnham . . . occasioning agreeable associations in the mind, and remembrances of the days of my youth, when I once resided in the town."

If at school at Farnham, it was probably at a dame's school, but he may possibly have been at the grammar school there, which was founded in 1611. He was certainly sent afterwards, though at what date is uncertain, to Basingstoke Grammar School,* the headmaster of which, the Vicar, was the Rev. Thomas Warton, father of two sons—Joseph, who

* The building, now a ruin, was used as a school up to 1855.

matriculated at Oriel College in the same year as Gilbert White, and Thomas. They became men of some literary eminence, and were respectively headmaster of Winchester College; and Professor of Poetry at Oxford and Poet Laureate. Neither of them, however, can have been for any length of time school-fellows of the Selborne naturalist, since Joseph Warton went to Winchester in 1736, and Thomas was born in 1728, and therefore eight years the junior of Gilbert White.

One incident of his school days at Basingstoke is mentioned in "The Antiquities of Selborne," Letter XXVI.

"When a schoolboy, more than fifty years ago, he [the author] was eye-witness, perhaps a party concerned, in the undermining a portion of that fine old ruin at the north end of Basingstoke town, well known by the name of *Holy Ghost Chapel*. Very providentially the vast fragment which these thoughtless little engineers endeavoured to sap did not give way so soon as might have been expected; but it fell the night following, and with such violence that it shook the very ground, and, awakening the inhabitants of the neighbouring cottages, made them start up in their beds as if they had felt an earthquake. The motive for this dangerous attempt does not appear: perhaps the more danger the more honour, thought the boys; and the notion of doing some mischief gave a zest to the enterprize. As Dryden says upon another occasion,

"'It look'd so like a sin it pleas'd the more.'"

In his first letter to Pennant (which, from the insertion of preliminary matter when his book

CHAPEL AND SCHOOL OF THE GUILD OF THE HOLY GHOST, BASINGSTOKE

[To face p. 30, Vol. I.

was prepared for publication, became Letter X.) Gilbert White states that he had been from his childhood attached to the study of natural information. Confirmation of this appears in a MS. diary kept by Thomas Barker, of Lyndon Hall, Rutland, who subsequently became his brother-in-law:—

"1736.—March 31. A flock of wild Geese flew N.—G. W.
„ April 6. The cuckow heard.—G. W."

The initials certainly refer to an observer, and are doubtless those of Gilbert White, who, then a boy of fifteen, was probably staying with his uncle and aunt, Mr. and Mrs. Isaac, at Whitwell Rectory, near Lyndon—perhaps on an Easter holiday. Writing in his seventy-first year to Robert Marsham, he mentions that he himself had been "an early planter"; since, in 1731, as a boy of eleven years, he had planted an oak and an ash tree in his father's grounds at Selborne.

Throughout his life it was Gilbert White's methodical habit to make regular entries of his expenses, receipts, and other items; and accordingly the following entry occurs in his pocket-book, together with "an account of my cloaths," of the same date:—

"An Account of Books that I carryed to Basingstoke January 17, 173$\frac{8}{9}$.

"Greek Testament bound with the Common prayer.
Whole Duty of Man.

Latin Grammar.
Two Greek Grammars.
Walker's Particles in two Vollumes.
Hill's Lexicon.
Terence.
Erasmus.
Cornelius Nepos.
Bible.
Sallust.
Castalio's Latin Testament.
Xenophon de Cyri Insti:
Sallust Delphin.
Tully's Epistles Ditto
Horace Ditto
} All these to be returned to Uncle White when I leave Basingstoke.*
Virgil Delphin.
Homer.
Ainsworth's Dictionary.
Isocrates.
Tompson's Seasons.
Scholar's Manual.
Practical Greek Grammar.
Tacitus.
A Translation of Horace.
Gradus.
Horace (little).
Cæsar's Commentaries.
Wilkins's Natural Religion.

* From this sentence, and indeed from the occurrence of the whole list of books at this time, it might almost be inferred that this was the date of the first admission to Basingstoke Grammar School, notwithstanding the statement of his nephew John White (who was not always as exact as he might have been in his remarks) in his "Advertisement" to the second edition of his uncle's book (2 vols. 8vo, 1802), that the naturalist "received his school education at Basingstoke." If this surmise is correct, Farnham Grammar School is probably responsible for the chief portion of his education, but since no records of either school are extant the matter is unlikely to be ever cleared up.

Tullie's Orations Delph.
Horace, Minel.
Tully de Oratore Delph.
Introduction to the Lord's Supper."

A list which would seem to indicate a respectable amount of scholarship in a boy of eighteen.

On December 17th, 1739, he was admitted a Commoner of Oriel College, Oxford, but does not seem to have resided till April, 1740, when certain "cloaths," in the shape of sheets, etc., were sent to Oxford. On November 18th, 1740, he paid, however, university fees "for a whole year to November 18," and at this date he paid his "caution money" (£8), bought a gown, paid £4 5*s.* 7*d.* for "thirds of my room," *i.e.* for its furniture, which he supplemented with "chairs of Mr. Bentham," his tutor No record is kept of the rooms occupied by undergraduates; but probably his rooms were in the front quadrangle at Oriel, since there is mention of the letting of "the little room belonging to my chamber to Mr. Andrews servitor for 10*s.* a year," the rooms in the inner quadrangle not having two bedrooms. There are numerous entries of purchases of books, some of them being of a religious character; but entries of the purchase of gun flints, a "shot-charger and powderhorn," "pair of spurs," "a day's ride," "to the music club," and on one occasion "boat hire," show that his life was not an absolutely ascetic one.

Writing to him many years after this time, the Rev. John Mulso—his contemporary at Oriel and lifelong intimate friend—says :—

"I do not ever remember your shooting a snipe at Oxford in summer, where there used to be plenty in winter: at that time you used to practise with your gun in summer to steady your hand for winter, and inhospitably fetch down our visitants the birds of passage."

During the long vacation of 1742, a three months' visit was paid to his relations at Whitwell Rectory, Rutland. Time passed, and on June 13th, 1743, we find him taking out a "liceat for examination," of course for his B.A. degree.

This examination in those days, and long afterwards, was not quite so elaborate as it now is. In those fortunate times Gilbert White would have appeared before two Masters of Arts of his own college, and suffered at their hands but a very brief examination, which, if not entirely, was probably chiefly oral. On June 17th he is able to make the entry in his pocket-book :—

"For a Testamur and ringing the Bells
for examination . . . 1^s 6^d"

and on the same day he writes his name in the Vice-Chancellor's book (and pays a shilling). On June 22nd occurs this entry :—

"Proctor's men and major . . 1^s 6^d"

University men may perhaps recollect occasions on which their payments to the Proctor were not of a

1.

£ s d

Decem^r 6. 1743. Received then of Mr Benjef in Cash: 8 16 6
which with 10 = 0 = 0 in cash received last Octob: 4 = 8 = 8 Land-tax; & 4 = 8 offerings; with 1 10 2 Repairs & postage, makes 24 in full for a years Rent to Mich: 1743.

Decem^r 9. 1743 Received of Combs Half year's Rent to Midsum^r 1743 — — 3 0 0

N: B: Landtax bill was 1 : 3 : 4.

An Account of Expences at Oxford from Novem^r 9. 17

	£	s
Three pounds of Candles	0	1
Horses & Servant at the Anchor	0	
2 Ounces of green-tea — —	0	
Novem^r 24. Paid a Dinter's		
— Hire of a screen — — —		
Paid also Taylor's Bill —		
Gallon of Wine —		
Novem^r 25. Paid then half years Room-rent to Mich: 1743.	2	
D^o Paid two Quart^{rs} Battles to Mich: 1743. — —	8	
A Gimblet, Screws, & Carpenter's — Jobbs ————	0	1
Gun-flints — —	0	0
	11	17

2.

1. CUP presented to Sir Sampson White at the Coronation of Charles II.
2. POCKET ACCOUNT-BOOK used by Gilbert White at Oxford

[To face p. 34, Vol I.

quite voluntary character, but this entry probably merely refers to a customary fee, as does the following:—

"June 27th major and scout . . 3^{s} 0^{d}"

On the same day he pays the fees for his degree, and on the following day he was evidently in need of recreation, for he expends 4*s.* on "Horse hire and Dinner," and the same sum on wine.

Then he quitted residence for Selborne. He kept the next Michaelmas Term as a Bachelor, bringing up his gun and a dog, and continued his residence during the following Lent Term, when he pays fees to the Dean and the Beadle "for Determining." He attends "Dr. Bradley's first course of Mathematical Lectures," and on March 30th, 1744, he pays fees to the Vice-Chancellor's servant and to the ringers at S. Mary's on being elected a Fellow of Oriel. His actual election, however, occurred shortly before this date, and therefore in 1743 O.S.

That Gilbert White was a man of sound scholarship and wide reading needs no proof—his writings abundantly show it. Nevertheless, it has been a little the fashion to assume that but a very small modicum of learning was demanded of candidates for fellowships in his day. That other things besides scholarship were considered, must be admitted; nevertheless an examination was held. By a statute made by Oriel College in 1722, it was laid down that the election of fellows was to take place on

Friday after Easter: "Candidates to appear not later than the Wednesday before, candidates to be examined on the Thursday, and the election to be completed before 6 o'clock p.m. on the Friday."

Thomas Hearn, the Oxford antiquary, in his diary of May 27th, 1709, writes:—

"This day was an election of fellows at Oriel Coll. There were three vacancies and nine candidates for them, there were two of Oriel Coll. who stood, and one of them came in as being pupil to one of those chiefly concerned in the election. The second was of Merton and the third of Wadham Coll. Mr. Johnson an ingenious* good-natured modest Gent. of Ch. Ch. stood and performed better, at least as well as any; but interest swayed (notwithstanding what was given out both before the election and since) as I have been informed by one of the College, an observer of the transaction but perfectly unprejudiced (as having nothing to do in the election one way or other) and one of the electors has himself declared that he was engaged some-time before the time of trial by a gentleman in the country. So that both in this as well as other colleges things are managed by interest, not by merit."

Evidently interest was not altogether absent, but the above extract at least shows that there was a *trial*, *i.e.* an examination, and it should be noted that at this time Oriel fellowships were open to candidates from other colleges, which was then a quite exceptional practice.

Dr. Shadwell, writing the account of Oriel in

* *i.e.* well-born.

Clark's 'Colleges of Oxford,'* states that "the election to fellowships was singularly free from restriction: for most of them there was no limitation of birth, locality, or kindred, and no class of junior members had any title to preference." So that it may be reasonably concluded that something considerably greater than a slight modicum of learning was required, even at this time, in the new Fellow of Oriel; who proved to be in after-life a man of scholarly tastes, keeping up his classical knowledge, while he did not permit its claims to exclude the study of history and travel, to say nothing of his devotion to natural history.

During his probationary year Gilbert White, of course, resided at Oxford, and there he spent the summer of 1744. From this time to the year 1791, two years before Gilbert White's death, a complete series of letters from an Oriel contemporary has been preserved and will be freely quoted from. This correspondent, who continued all his life on terms of sincere friendship with Gilbert White, was John Mulso,† the second son of Thomas Mulso, of Twywell, Northamptonshire, who lived, however, chiefly in London, the representative of a family long settled in that county. Born in 1721, John Mulso was educated at Winchester College, which he left third on the roll in the same year (1740) as Collins the

* p. 121.

† Unfortunately the letters from Gilbert White were destroyed after the death of John Mulso.

poet, who was first, and Joseph Warton, who was second. Apparently he first made Gilbert White's acquaintance at Oriel College, where Mulso was slightly the junior in standing.

Mulso frequently mentions his eldest brother, Thomas, who married a Miss Prescott on the same day (in 1760) that his only sister Hester married a friend of Richardson the novelist, one Chapone, an attorney practising in the Temple. Another brother, Edward (Ned) Mulso, who died in 1782 unmarried, is also often mentioned. With all the members of this family Gilbert White was on terms of affectionate intimacy.

Writing from Leeds Abbey, near Maidstone, in Kent, on July 18th, 1744, John Mulso says—

DEAR GIL,—I suppose that by this time you are returned to Oxford to prove to the Orielenses how worthy you are to be a Fellow, how compatible that title is with the character of the gentleman, that without Formality Respect may be preserved, and that to depend does not always signifie to be servile. I long to hear from you and to know the state of that poor college, which I do not expect to see again these many months. Let me know what sort of liberties you are allowed, and who remains to share them with you and make them sweeter. . . . I saw Collins * in town, he is entirely an author, and hardly speaks out of rule. I hope his subscriptions go on well in Oxford. . . . My mother loves you, you have a strong party in a family that you never saw, but I claim your heart, and am, with sincerity,

Your affectionate friend and humble servant,

J. MULSO.

* The poet.

In August, 1744, the same correspondent wr'tes that he had for the first time seen the sea at Hearn—

"A sight which I have often been obliged by you with a description of, and now find the justness of your description. . . . The Ladies at Canterbury are insufferably handsome, I never met in one place such an assembly of Beauties, I believe. I saved my heart by the beautiful confusion. I could justly say 'defendit numerus'; I was vastly amused at your account of your danger in that way. . . . I have seen some pretty good pictures since I have been out, but why do I mention pictures to you who have seen Burleigh. . . . My mother loves you so much that I am almost jealous of you. She says you speak her very sentiment in your judgment of the Odyssey. She thanks you for your advice to me, and I suppose now I am returned to Leeds [Abbey] I shall be forced to mount pretty often, and desert my dear corner. I don't find any great propensity to poetry. . . . I have not seen a verse since I left you, nor hardly heard one quoted, so that I am not incited to it. I believe the return to this dull place would have made me hang myself, if we had not brought back with us a little company from Canterbury: like you and Falstaff, I hate compulsion, and I am sure I stay here against my will."

On September 4th he writes again, exhorting his friend to come and visit his family in London. He continues—

"I suppose Treufle is in high favour now; he is in season, and by this time has beat round all the manor where you obtained leave to shoot. . . . Tell me how you all go on at Oxford. Tell me if Jenny has resumed her Empire, and totally expelled her rival the Stamfordian. Tell me if she shall be your Penelope, for since your reading the Odyssey, I suppose you have learned to despise the whip-syllabub names of Amoret and Saccharyssa."

Towards the end of the month Gilbert White rode home to Selborne (with his gun in a leathern bucket).

On October 8th, 1744, Mulso writes—

"... Do you really find celibacy hang heavy on your hands, or does Tom only jest when he says that you are on the high road to the dreary and dolorous land of matrimony? Upon my word, I would not advise you to play so much as you do with the tangles of Neæra's hair: these meshes will hold fast a heart of stronger wing for flight than yours is, and if

'Beauty draws us with a single hair,'

it may very well hold with a whole tête. . . . Collins is now my next neighbour. I have breakfasted with him this morning."

On November 6th, 1744, Mulso writes from London—

"I turn my eye towards Selbourne. I long to see an old friend in a new place. But alas! journeys are too expensive for a younger brother."

In February, 174$\frac{4}{5}$, residence at Oxford was resumed. On June 26th the entry occurs—

"Threw up my room to Mr. Barnes."

July 17th, however, found him back at Oxford, but he subsequently went home to Selborne; and in September to pay one of those visits at Ringmer, near Lewes, to his aunt Mrs. Snooke, the owner of Timothy, "the old family tortoise," which became of almost annual occurrence.

Writing to him here on September 7th Mulso says—

"I thank you for letting me know of your Pleasures, it is always an addition to my own. I never, as you know, read the Odyssey."

Referring to an invitation to Selborne, he continues—

"I look for an Arcady with you, and I expect some kind whispers from the unseen genius of your woods. I have seen no pictures since I was at Windsor with you. . . . Collins has been some time returned from Flanders, in order to put on the gown as I hear, and get a chaplaincy in a regiment. Don't laugh. Indeed I don't on these occasions. This will be the second acquaintance of mine who becomes the thing he most derides. . . . Heck [his sister] likes your hair. She confesses so much already. It was a very neat compliment that you sent her. She can't answer it, so she says nothing."

On October 23rd, 1745, Mulso wrote of a visit with his father to Northamptonshire—

"I saw a good deal of the country. I am sure you would like it; as we walked the fields we sprang large coveys of partridges and started hares. I should be glad of a large hunting seat there for your service; I call my father to witness how often I repeated 'if White was here how happy he'd be.'"

Early in the following year, $174\frac{5}{6}$, an event of considerable importance to the White family occurred. Mr. Thomas Holt, receiver to the Duke of Bedford, with whom he was connected through the Howland family, died at Thorney in the Isle of Ely. Mr. Holt, who was a half-brother of Gilbert

White's maternal grandfather, by his will entailed property in Essex upon Thomas White, the next eldest brother to Gilbert White, subject however to certain annuities, which, as it happened, precluded him from receiving much benefit from the estates until, many years afterwards, in 1776, the chief annuitant died. A small amount of money, some three hundred pounds each, was inherited by the other brothers.

Gilbert White attended the sick-bed at Thorney, and, being one of the executors and trustees, remained for some time to wind up Mr. Holt's affairs at Gore Farm, of which he was tenant under the Duke of Bedford. In the intervals of his business cares he seems to have indulged himself in a practice which remained with him for life—that of writing verses, which were sent to his friends. And at this time he composed the first version of his little poem, 'The Invitation to Selborne,' which charmingly describes the beauties of that place and neighbourhood.*

Mulso writes to him on February 9th, $174\frac{5}{6}$, and after what may be called "congratulatory condolences" upon the loss of a relation, continues—

"I fancy you will hardly leave the parts you are now in, without making a visit to Stamford,† and then I suppose

* cf. Letter from Mulso of September, 1758, in which he mentions having received the 'Invitation' originally "out of the Fens of Cambridgeshire."

† Probably to a family named Brown there. Mrs. Barker's eldest daughter, b. 1752, married a Mr. Edward Brown, of Stamford.

will cross away to Hampshire, where your brother Benjamin told me you was expected. . . . I am obliged to you for putting the finishing stroke to your poetical Performance of which you gave me a Hint whilst you were in Town. I can easily discover which were breath'd out in the pure air of Selborne, and which in the Fens; tho' I don't mean that they are tainted, for the conclusion is as charming as the beginning. I fancy my sister Sung herself into Six Lines, when you was in town last. I assure you without flattery they are very much and very justly admired. . . . If you pass through London on your return don't neglect King's Square Court [his father's residence] tho' I am not there. You will find many admirers, who lov'd you first for my sake and now have a stronger reason to do so."

Later on, Gilbert White went to examine Mr. Holt's Essex estates, and wrote from one of the farms near Rochford to his father on April 9th, 1746 :—

HONOURED SIR,—Finding by your letter to me, that Franky* brought to London, that you was inclined to have me go to this part of the Country, I set out, and arrived here yesterday.

After some details concerning the property and tenants, the letter ends with a significant description of the state of Essex roads at this time—

I think the Hundreds in general look pretty pleasant; only the fear of agues makes people dislike them. The Roads in the Vales are bad, and there you meet with a horse-track that has a sound bottom, which is very convenient for Horses that know how to go in it: but my

* His brother Francis.

horse found it so narrow that he was much foiled, and has cut himself all four. I hope the short time I stay here, I shall escape the ague.

I am, Sir,
Your dutiful son,
GIL WHITE.

On April 16th, 1746, he writes again to his father from London. The letter concludes—

I did to the best of my power survey all the Estates round Rochford, and took my remarks down in my pocket-book on the spot, and will send you a copy of my observations from Thorney. My Sister sets out for Ringmer to-day behind Daniel Crips. My brothers* are both pretty well. Franky has got a cough.

I am, Sir,
Your dutiful son,
G. WHITE.

P.S.—I received your Letter of the 15, but have no time to answer it this post. I burnt the first page.

The elaborate and tabulated observations are extant; and the writer of the present memoir, who knows the property very well, is able to testify to the great care and accuracy with which they were compiled.

On April 23rd, 1746, he writes again to his father from "Gore Farm," Thorney, Isle of Ely—

"After a long variety of journeys, I thank God, I arrived safe here this evening, and found my Brother† very well.

* Benjamin and Francis, who had commenced their business career in London.

† Thomas, who writes a short note to his father on the same letter-sheet.

I made Harlow* in my way hither, and took my observations in my pocket-book as usual; which I will send you by Letter when I have more time. . . ."

The following memorandum, which was made during his residence at Thorney, well illustrates his careful and methodical habit of mind:—

"To sell the sheep as fast as possible. As many oxen as are saleable. Not to sell the Plate by auction at Thorney, but to reserve it to be disposed of at London by weight. The four men-servants not to be discharged 'till the will is proved, because they are witnesses. To take great care of the papers in the 'scrutore in the best Chamber, especially Bonds, Ledger, &c. Use great secresy about money matters."

On May 28th, 1746, John Mulso addressed a letter to his friend at Selborne, thanking him for entertainment there, and begging him to write "before and during your excursions." For the man, who has of late years been frequently represented as a stay-at-home recluse, was off again in the following month (June) to Spalding, in Lincolnshire, where he "was visiting for a week together."† Probably he went there from Thorney, since in a letter to Marsham, written in 1791, he states positively that in the year 1746 he "lived for six months at Thorney, in the Isle of Ely, to settle an executorship and dispose of live stock."

* In Essex, near which town another manor and farm of Mr. Holt's were situated.

† *Vide* 'The Natural History of Selborne,' Letter XXIII. to Pennant.

On August 1st, 1746, Mulso writes to Selborne—

"I have a little longer deferred writing to dear Gil, suspecting that the charms of Todnam* would occasion my letters lying unopened at Selbourne if I wrote sooner, and I cannot believe you passed by Tom Mander so quiet as you told me you would: if I know him he has the Art of engaging a little longer, and yet a little longer tarrying. Lucretius's 'suave mari magno,' etc., was not the reason I laughed so heartily at your stage-coach sickness, which you have now recovered. I hope you will forgive me. I believe it was rather the circumstances of the sickness than the sickness itself, that diverted me. I don't think there is a better answer to the Question of Original Sin than a groan; or a better satire on woman's disputing it, than your cascading."

This infirmity was no laughing matter, however, to his friend; since his sufferings from it prevented him, as he told Pennant (in an unpublished letter), from making the long journey to Flintshire on a visit to Downing, Pennant's seat there. Perhaps this was the reason why he so constantly rode everywhere, the names "hussar parson" and "a centaur not fabulous" being bestowed on him by Mulso.

The letter continues—

"I have just received a letter from Collins, dated Antwerp. He gives a very descriptive journal of his travels through Holland to that place, which he is in raptures about, and promises a more particular account of. He is in high spirits, though near the French. He was just setting out for the Army, which he says are in a poor way. . . . I dare not tell you how much we think ourselves obliged to you for your company in town."

* Todenham, near Moreton-in-the-Marsh, Gloucestershire, where lived a college friend, Thomas Mander, who became a Fellow of Oriel.

August 9th, 1746, found him again up at Oxford, when he attends the races in September. On October 7th a "liceat for declamation" costs 6*s.*, and on October 22nd he pays "fees at taking A.M. degree 3^{l} $6^{s.}$ $8^{d.}$"

A somewhat higher scale of fees for the M.A. degree is now in vogue at Oxford, but no sort of examination. At this date, however, it was not so. John Mulso, writing on October 27th, 1746, says—

"I wish you joy of having passed the fiery ordeal of M.A., but I am sorry you ended so furiously as to burn your works; why was not Augustus at your elbow to rescue those unfortunate compositions? To say truth I should have been glad to have seen them, for tho' I might not copy, I might imitate, and I want a model, and for that want's sake I defer setting about anything of that nature; tell me your whole Proceeding, tell me your Questions, tell me your Theses, tell me your examination, your masters, your—— let me into the whole Fund. I desire you would always have wet brown paper about you, that is, I desire you would not mislay or carry away, or lend out, or in any other way distrain the Scheme, which upon your Promise I now call mine. I should be glad to have it here if you could contrive it."

He also mentions with approval a translation from *Horace*, Od. III. 26, which his friend had sent to him.

On Sunday, April 27th, 1747, Gilbert White received Deacon's Orders in Christ Church Cathedral, Oxford, from Thomas Secker, Bishop of Oxford. He at once became curate to his uncle (by marriage), the Rev. Charles White, Rector of Bradley and Swarraton, Hants, with duty at Swarraton. Notwithstanding the fact of this curacy—which, however, seeing that his stipend was only £20 a year, probably only entailed Sunday duty—he came up to Oriel again on July 11th; whence, as he kept a mare at this time at livery, he seems to have continued his weekly attendances at Swarraton.

During the summer he again visited his college friend, Tom Mander, at Todenham, in Gloucestershire, where Mulso addresses him on August 21st, 1747—

"I presume you are popping and snapping so that a Farmer can't walk his own fields in security for you. Tom can walk farthest, but you shoot best; I fancy I have drawn your characters, tho' I may add Tom drinks cyder longest, but you take the larger glasses at first. I thank you for your account of yourself at Chalgrave. There is no man understands a Retreat I see better than yourself."

Miller's 'Gardening Dictionary,' apparently the first book he possessed which relates to his favourite studies, was purchased during this year.

On October 16th, 1747, Gilbert White was laid up at Oriel with an attack of small-pox. Some of the entries found in his pocket-book under the heading, "Expenses in the Small-pox," seem to imply a different treatment of the disorder from that now adopted, and are worth reproduction.

"October 16th—

	Li.	*s.*	*d.*
3 bottles of white wine . .	0	6	0
Bottle of wine . .	0	2	0
Half a pound of Corinths .	0	0	3
Paid Jo for watching . .	0	1	6
Ounce of green tea . .	0	1	1½
2 pounds of sugar . .	0	1	8
Figs and Corinths . .	0	0	7
Pint of wine . . .	0	1	0
Ounces of Bohea . .	0	0	6
Pound of Rushlights . .	0	0	6
Bohea Tea . . .	0	1	0
Dish of Tripe . . .	0	0	6
Ounces of Green Tea . .	0	1	0
Coffee-house for Balm Tea .	0	1	9"

Judging by the amount of the fees paid to two doctors, together £31 10*s.* 0*d.*, and from the fact that a nurse was sent from Hampshire, the attack must have been a rather severe one. The patient seems, however, to have quite recovered by the end of December, if it was for his own use that he then purchased skates and a shooting net.

It has been suggested that the reason Gilbert White never sat for his portrait was that he was badly marked in the face as a result of this attack of small-pox. Whether he was permanently so marked or not, it is now impossible to say. But no existing member of his family seems to have heard that he was disfigured. The remark by Mr. Bell,* that

* *Vide* Bell's edition of 'The Natural History and Antiquities of Selborne,' vol. i. p. lviii.

"he never would sit for his portrait," is not correct. The present writer has often heard his father, the late Algernon Holt-White, say, upon the authority of his own father, Thomas Holt-White, who was Gilbert White's nephew and knew him very well, that the naturalist showed no disinclination to have his portrait painted when urged to this course by his brother Thomas, but that he did not, as it happened, ever have it done.

Writing to his friend Mr. Churton many years after this time, on August 20th, 1783, Gilbert White mentions a copy of verses as appearing in 'The Gentleman's Magazine' for June, 1783, "written by a poor dear Oxford friend long since dead, about 35 years ago," *i.e.* in 1748, soon after this attack of small-pox.

From these verses, which are here appended, it will be seen that there is what was perhaps a merely jocular mention of the "roughen'd face," but it hardly follows from this that he was badly disfigured in later life, or indeed marked at all.

"THE METAMORPHOSIS

"Corycius long admired (a curious swain!)
The wealth and beauties of Pomona's reign;
The vegetable world engrossed his heart,
His garden lingering nature help'd by art;
Where in the smoking beds high heap'd appear
Sallads and mushrooms thro' the various year.
But of each species sprung from seed or root,
The swelling melon was his favourite fruit;

Other productions kindled some delight
In his fond soul, but here he doted quite.
When others wisely to the grot retreat,
And seek a friendly shelter from the heat,
Anxious and stooping o'er his treasure, low
Poring he kneels, and thinks he sees it grow.
One day when Phœbus scorch'd the gaping plain,
Striving to rise at length he strove in vain,
Fix'd to the spot, exchang'd his shape and name,
A melon turned and what he view'd became.
Ovid would tell you how his roughen'd face *
Retains the network and the fretty grace;
His skin and bones compose the tougher rind;
His flesh compressed retains its name and kind;
Shrunk are his veins, and empty'd of their blood,
Which in the centre forms a plenteous flood.
The morning past away; 'twas noon; 'twas night;
'Twas noon again; no lord return'd; their fright
The servants own'd; when one cry'd out 'I've found
'The secret now, he's in the melon ground,'
And straight ran thither, then he call'd amain,
The adjacent hills re-echoed to the strain;
But as he look'd about, ripe at his foot
A melon lay, just waiting to be cut:
He urg'd the fatal knife:—when burst a groan
With words like these, 'you've stabbed your master, John.'
So bleeding twigs the Trojan hero tore,
And hollow murmurs shook the Thracian shore." †

"* By the small-pox." "† Æn. iii. 40."

During 1748 the Oxford residence continued to the end of April, when he let his rooms "for a quarter of a year," a transaction which he repeated in July for another quarter. Then he quitted residence at the University and went to Hampshire, where he continued his duties as curate of Swarraton, near Alresford. He seems to have chiefly divided his time between Bradley, where his uncle, who held that living, resided, and Selborne. At this time entries respecting his gun and dogs appear in his account-book, and in May, 1749, "a pair of hunting boots" was purchased, which, judging from the date, could have scarcely been intended for riding to hounds. Visits were paid to Midhurst and Cowdray House, which he showed to his youngest brother, Harry, at this time a boy sixteen years of age. In June a visit was paid to John Mulso at Sunbury Vicarage; and to Ringmer, whence he returned in the middle of August. A little later on a journey "to Kingston with Harry," who was at school there, was made. In September, Winchester and Portsmouth were visited with Mulso.

It is pleasant to notice that his account-book at this time contains constant entries of gifts to the poor. One of the entries is:—

"Gave Goody Marshall pair of shoes for nursing me in the small-pox."

CHAPTER IV.

ON March 11th, 1749, Gilbert White was ordained priest in the Chapel in Spring Gardens, London, by James Beauclerk, Bishop of Hereford, acting for the Bishop of Winchester. At the following Easter, when, as usual, he went to attend the Oriel College audit, he took his brother Harry to enter him there, though the latter did not commence residence until the following year. At this time a purchase was made of a pony, "Mouse," which was destined to carry its owner for many years about the lanes and over the downs of Hampshire and Sussex.

Though Gilbert White could be liberal to his poorer neighbours, he commenced the habit of living well within his income very early in life. The following entry occurs on October 27th, 1749:—

> "Purchased then with £300 legacy money and £50 Saved £333 12^{s} 10^{d} at $104\frac{3}{4}$ in Bank Annuities at 4 per cent."

This careful habit of saving, when he could, was retained during all his life.

On July 17th, 1749, John Mulso writes from Sunbury, after a visit from his friend:—

" . . . You made my Sundays Sabbath days indeed and all the week Festival. We retain several of your expressions, and are pleased to fall into your Manner. You steal our songs and we your sayings: in short we are never more pleas'd than when we can set you before us. . . . you have the art to be truly companionable. Miss Hecky has been a Rake and deserted us for two whole days, and went to the Races and Assembly and danced away in company with Lady Musgrave: I wish you was here to hear her description of Races: the *Sophoclean* Ὤμόι, woe's me! was used with energy."

The future "Admirable Mrs. Chapone" on a race-course suggests indeed a curious picture!

He goes on to describe the proposed journey of a family party to Oxford, and continues—

"Heck is in the greatest alarm, and screamed out on hearing it 'but where's my Busser?'* In short she is apprehensive of a dearth of civilities, because you are not to be there; and fears she shall not get her Degree, because she has not her favourite Batchelour to answer under."

The "Heck" here mentioned was his sister, Hester Mulso (b. 1727, d. 1801).

In Jesse's edition of 'The Natural History of Selborne'† the statement is made that "Mr. White in his earlier days was much attached to Miss Mulso (afterwards Mrs. Chapone)," and Mr. Bell‡ has ex-

* A nickname which Gilbert White received at Oriel.

† *Vide* "The Natural History of Selborne," ed. Jesse, 1854, p. xi.

‡ *Vide op. cit.* Bell's edition, vol. i. p. xxxv.

panded this statement into a touching little romance, in which Gilbert White figures as a blighted being, who never married because of the enduring effect of his early disappointment. But all this rests upon no foundation whatever in fact. Mr. Jesse, who certainly saw some of Gilbert White's unpublished MSS., had very probably also been shown the letters to him from John Mulso (which, indeed, he mentions in the next paragraph), and had noted cursorily some such passages as the one above quoted. Anyone who examines the whole series of these letters with the slightest care, however, will see at once that, though John Mulso and his family, who were really very fond of Gilbert White, would have welcomed the union, neither the gentleman nor the lady, though excellent and intimate friends, ever appear to have manifested the least hint of anything more than friendship for each other. 'The Posthumous Works of Mrs. Chapone, etc.,'* contain "An Account of her Life and Character, drawn up by one of the Family." Herein there is no mention of Gilbert White. nor even the slightest allusion to him, though her liking for Mr. Chapone "from their first introduction," and their three years' engagement previous to marriage in 1760, are dwelt upon. Nor is any tradition of the disappointed affections known among the family of the naturalist, who had but one mistress—Selborne.

* John Murray, 1807 (2nd edition).

On September 25th, 1749, Mulso writes :—

"I very heartily thank you and your good family for all their favours at Selborne and Bradley, and I am very sorry that I could not make one more visit at the last place."

Judging from the address on Mulso's letters, Gilbert White was at this time living between his uncle's house at Bradley and his home in Selborne. There are now frequent references to purchases for the garden at "The Wakes," and on October 13th, 1749, he pays 1*s.* to

"John Breckhurst for planting trees on Selborne plaister,"

a practice in which he often indulged in later days.

On October 25th, 1749, Mulso writes describing a visit to "the top of Cowper's Hill." He continues :—

"We had just time to get to Egham to dinner, where Hunger recommended a plain meal, but we compensated that by the elegance of our entertainment, which was the reading of your Letter, and which I had hoarded up for that Purpose. I shall only say upon our expedition, that I thought the scenes it afforded us well worth going to see, even after the many beautiful ones which you took such kind pains to show me. . . .

"Are you sure that you shall not fetch home Harry at Xmas? Cannot you rest at Sunbury for a day or two? I think the sight of you would give me spirits for a month at least. If you time your visit well, perhaps I may take a trip to London with you; we will see the family; salute the Bishop, and I wish with all my heart he could give you a Hampshire living of two hundred a year in return for the compliment. Perhaps we may steal into the pit and take another laugh together, unless you think it beneath your dignity. But you shall rule; and indeed this is a compliment which I owe to you, who referred everything to my arbitration, and submitted everything to my content and pleasure when I was in Hampshire. . . . Tell me about the curacy and whether you have taken it."

From the last sentence and an entry in his account-book of the receipt of a year's salary (£20) for serving the curacy of Swarraton to March 21st, $174\frac{8}{9}$, it would seem that Gilbert White had been without regular duty for some months.

The bishop mentioned in this letter was Dr. Thomas, Bishop of Peterborough, and subsequently of Salisbury and of Winchester, who had been tutor to the Prince of Wales (afterwards George III.) and his brother Prince Edward, who died young. He had married Mulso's aunt, Susanna. Though he never gave Gilbert White a living, he was, as will be seen later, a remarkably benevolent uncle to John Mulso.

On January 24th, $174\frac{9}{50}$, Mulso writes:—

"I thank you for entrusting me with Horace; I have long expected his coming, as I overheard several hints from him the last time I saw you. . . . Wherever I have carried him it is agreed that he is as well dressed, and presents that easy wit and humour, which he exhibited when Mr. Pope brought him and introduced him to the town some years ago. The Bishop of Peterborough, who remembers him in his Roman dress, and indeed before he could speak English, thinks it wonderful that he should be so well reconciled to our language and manners. He is perfectly naturalized. But to leave Personification and to speak of your Piece; I like it very much and so does everybody to whom I have show'd it; only Miss Hecky likes more that you should be indebted to nobody for your plan. . . . I am glad that you have such an inclination to Sunbury, I wish the word *Præbeo* was more applicable to me, that I might accommodate you in the amplest manner. You may well call the vicarage a Caravansary, for as I take it there was little more than bare walls in them. . . . I call Mouse to witness my Eastern beggary. You must settle your own time and then I will not be in London, but I will attend you thither with pleasure. Miss Hecky will be glad to see you, and so will Jenny* etc. I expect to see your brother Musgrave† soon who is going up to attend his sister Betty's marriage. The day is not fixed. Dont rail at my laziness. I have in some of your letters the best descriptions of Procrastination, and applied to yourself. I profess it, my genius is idle, but I dont glory in it neither. . . . You may venture to a Play, for the audience had like to have torn the House down the other night, for the manager's daring to revive an indecent Play of Otway's; and calling him forth, ordered him never to affront their ears with such loose performances."

On April 11th, 1750, John Mulso writes again :—

"You are now I suppose to be found, like Cyrus, ranging your Trees, and nursing your Plants. . . . I wish you joy of the arrival of the swallows and the swifts and the nightingales. I hope you will send me word how your nurseries go on, and the true state of Selbourne Hanger, with the delightful history of the Temple and Weathercock. . . . I hope you are writing out Heck's sermon‡ for us, for the Bishop is soon going down to Peterboro', and

* Miss Jane Young, to whom Mulso was engaged.

† Chardin Musgrave, Fellow, subsequently Provost, of Oriel.

‡ A sermon composed by Miss Mulso and sent to G. White.

He wants to have it. The Bishop loves you. . . . I hope the spring will invite out a little of your Poetry, you know what Heck said, that she would not have you always translate and imitate, but give your own Invention scope, and I hope you observe what she says."

A little later, in May, speaking of an excursion of his sister's into Kent, the same correspondent mentions that the Kentish road affords fine scenery, "as I believe you have experienced," and remarks that his sister said her "second day was according to Mr. White's taste a *cool, brown* day; but as she is as much given to Agues as you are to Fevers, she invok'd the blessed sun to come and warm her, with as much earnestness as you creep into the shades, or shelter in the 'nidus Acherontiæ,' by which name I think you have christened your arbour on the Hill."

Possessors of either of the quarto editions of Gilbert White's book will see in the large north-east view of Selborne this "arbour," or one which subsequently superseded it, called the "Hermitage," near the zigzag path (not yet made however) up the Hanger. The other little summer-house shown in the large north-east view of Selborne, together with the sloping path, or "Bostal," near which it stood, was not constructed for many years after the time now treated of. The latter was called by Gilbert White "the new Hermitage."*

* These two little buildings have of course long perished, but the platforms dug out of the chalk hill, on which they stood, are plainly to be seen. From the bostal, a little below where the later Hermitage stood, a small, but still perfectly-defined zigzag path leads down to a wicket-gate opening into the little park behind "The Wakes."

The construction of this arbour may seem a very trivial matter, but it was the beginning of many subsequent embellishments and improvements, which most certainly tended in no small degree to keep their contriver in his native place.

During the autumn of the year 1750 Gilbert White visited his relation, Francis White, D.D., Rector of Christian Malford, in Wiltshire, and went on to stay at East Allington, near Kingsbridge in Devonshire, with an Oriel contemporary, the Rev. Nathan Wells, who had married and settled there as Rector. In this neighbourhood the Selborne visitor spent some weeks. The district, known as the South Hams—the garden of Devonshire—lying between the Rivers Tamar and Teign, and bounded by Dartmoor and the Channel, is perhaps the most fertile and charming in the beautiful county of Devon; a land of shady dells watered by clear streams and traversed by labyrinthine lanes, from which the traveller can ascend to the misty solitudes of Dartmoor. No doubt it was thoroughly enjoyed by the naturalist, than whom, probably, few young men of his age and position enjoyed greater opportunities for travelling about their native country.

Addressing him "In the Depths of Devon," Mulso writes, on August 30th, 1750:—

"You live a scrambling, rantipole life, and have a great variety of objects to be painted upon paper, (at which Landscape painting I think you a great and masterly hand) and sent to your sedentary friends. We receive them and think we are travelling with you for five minutes, and then look up and find ourselves in the same tedious scene."

After describing the country near Rickmansworth, Mulso continues:—

"This unevenness, however, gives a great Beauty to the whole, and makes it much in your taste; indeed I never see a spot which lies much out of the level but I think of you, and say, 'now this would please White.' . . . I hope you will write a poem and call it *the Progress* describing your own rambles. It would make a fine Piece and might tempt gentlemen to examine their own country before they went abroad and brought home a genteel disgust at the thoughts of England. . . . My father sets out in a day or two to fetch home that great stranger Miss Hecky; she has a design upon the poor Vicar, and talks of careering about with 'Whit*ibus*,' who she says heightens and improves all parties; whether there is any particular Hint of improvement by the termination she is pleased to give to your name, you best know: as for me I never see those things, because I do as I would be done by; so you best know the meaning of your new name, and whether it is a fond abbreviation of your Oxford title of *Busser*. I must see you after your journies; why you are just at home at Sunbury! Miss Hecky must hear all your travels and you must lead her quick imagination through your *δεινὰ πελῶρα*, as Homer does, to your *περικαλλέας ἄγρους*."

In October following the desired description was evidently forwarded, for the same correspondent writes on the 6th of that month:—

"I am much obliged to you for your account of your travels, which was very exact, and very entertaining. If you would but continue your tours, and write to me from them, I should have material for a very useful and agreeable Pockett volume. . . . I follow you in imagination, and indeed, according to your description of the South-hams, it is the only way in which I can follow you thither."

In November, Harry White, who was going into residence at Oriel, was accompanied to Oxford by his eldest brother, who spent the term there. At this time he bought "a Pontipool snuff-box for aunt White," and "14 yards of stuff for sister Becky." A loss occurred, since his second, or "portmanteau" horse, as he terms it, died on the return journey to Selborne. Oxford was, however, losing its charm for the Fellow of Oriel. Writing to him there on December 13th, 1750, his friend Mulso observes:—

"I love the thoughts of Oxford, yet I agree with you that it has not the same charms for me that it had. . . . I suppose Harry is heir to your old *goût* for a college life."

After Christmas the quiet life at Selborne seems to have been resumed.

In January, 1751, the family circle at "The Wakes" lost another member; since Miss Anne White was married to a gentleman whom she had probably met at her uncle, Mr. Isaac's, rectory at Whitwell, Mr. Thomas Barker, of Lyndon Hall, Rutland; a member of a family long settled in that county, of which Sir Abel Barker, Bart., of Hambleton, was the head.

This year (1751) is remarkable for the inception of a kind of diary, consisting chiefly of daily events in the garden, which was continued year by year for the whole of the rest of Gilbert White's lifetime. Possibly a remarkable man, Dr. Steven Hales, celebrated as a writer of works on natural philosophy, may have suggested the keeping of this journal. In the correspondence which arose after the publication of 'The History of Selborne' between its author and Robert Marsham, the well-known Norfolk naturalist, Gilbert White mentions his father's and his own friendship with Dr. Hales, who was his neighbour, he says, for part of the year, residing at his rectory of Faringdon, and of whom he gives some account. Marsham, in his first letter, mentions that he had kept a journal for above fifty years, and that it was undertaken "by the advice of my most estimable friend, the late Dr. Hales." Very possibly, though Dr. Hales' talents were chiefly directed to branches of science other than the observation of nature out of doors, Gilbert White's mind may have been considerably influenced by his friend's philosophical pursuits, of which he gives some account in his letter to Marsham of February 25th, 1791. Dr. Hales held the Rectory of Faringdon from 1722 to his death, in 1761.

This journal, which is headed thus—

'The Garden Kalendar for the Year 1751,'

was commenced and for many years written on leaves of quarto letter-paper. It was begun as an almost daily chronicle of the weather, the sowing of the usual garden seeds, of work in the garden, and of the planting of trees and shrubs, and alterations about the premises at Selborne, which, during his whole life, its author delighted to improve and embellish. Later there are notices of experiments and natural history observations, of visits to friends, and occasionally of family events. With the exception of a very few entries by his father and brothers Thomas and John, and at a later date by his old servant Thomas Hoar, the journal is in the handwriting of Gilbert White.

Those who possess Mr. Bell's edition of 'The Natural History and Antiquities of Selborne' will find in vol. ii. pp. 348–59 a reproduction of part of the *Garden Kalendar* for the year 1759.*

On June 6th, 1751, Mulso writes to Oxford from "London, in Miss Hecky's dressing-room. She is reading 'Clarissa,' on the other hand lies your *Invitation.*" In his usual manner he proceeds, as candid friends will, to criticise; and his remarks show that if his subject was the poem known as the 'Invitation to Selborne,' it must have been much altered at a later date. Very probably it may have been, since these verses were sent later on to other friends—among others, to Miss Catharine Battie some years afterwards.† Whatever it was, it pleased his friend,

* Much the same kind of journal was kept by Henry White at Fyfield Rectory, amusing extracts from which have been recently printed in 'Notes on the Parishes of Fyfield (etc.), by the Rev. R. H. Clutterbuck, F.S.A., edited by E. D. Webb, F.S.A.' Bennett Brothers, Salisbury, 1898.

† Her copy contains thirty more lines than are given in the version first printed in the 3rd edition of the 'Selborne,' 1813, and is otherwise slightly different.

who concludes his criticism with some very laudatory remarks, adding, "It is indeed agreed by all, from the Bishop to the Vicar, to be a *very good Piece*." He proceeds:—

"I find by your Hesitation that I must love one Proctor, though to say Truth, as you know, I never feared one. I assure you I think of the office with great respect, and as I make no Doubt that you will execute it with great applause, I think it may be a serviceable Honour to you. I fling in my vote to your accepting it, tho' I had rather have you at Sunbury to set me again on Horseback, which I took leave of when I took leave of you."

Miss Hester Mulso writes at this time on a sheet of her brother's letter as follows:—

"I am so much obliged both to the Poet and the Friend in Mr. White's gallant and elegant Invitation that I cannot help telling him how much I am mortified that I cannot thank him in Person for his admirable Poem. Your description of Selborne has left nothing to 'the craving Imagination of Miss Hecky,' and it was kindly done to send me so lively a Picture, as I fear I am not to see the Original. It is no great compliment to say that I wish to accept your Invitation, as I write from this suffocating Town, where I am killed with Heat, and have no Voice or Strength. Here, however, I am likely to remain (if I can exist) the greatest part of the summer, with only the refreshing excursion of a day or two now and then to Mr. Richardson's * at Northend, to keep me alive. I shall gratify his Vanity and my own by showing him your Verses; and I think yours, if you have any, must taste the praises of a Richardson. Pray give my thanks and compliments to your father and sister

* The novelist.

for their part of the Invitation. I hope your father has not seen your more than Poetical Compliments, for if he has, he must not see me unless he has a turn for Poetry, and knows that a Poet must *give* the Perfection he does not *find*. When you next drink Tea in '*the Pensile nest-like Bower*,' pity

Your obliged humble servant,

YES PAPA!" *

In June, Gilbert White went, after his brief visit to Oxford, to Chipping Norton; and on July 1st set out on "a journey to Oxford and round the N.W. of England to Lyndon and Stamford," returning on August 24th. In September, with Tom Mulso and his own brothers John and Harry, he made a little excursion to Portsmouth, and in October paid a short visit to Oxford, where he bought "of Mr. De Chair of Oriel a genteel bay mare coming 6 years old called Flora about 14 hands and a half high," which he notes he sold for the same money.

Writing on July 25th, 1751, to his friend at Lyndon, Mulso says:—

"You are going to see new Places and make more observations. I expect you would make them to me. Consider, I have a pretty collection of your travels about England, and you must carry the set through. I may make it a useful work in time, and enable young men to travel with Taste and improve at home."

On October 25th, 1751, Gilbert White became Curate-in-charge of Selborne for Dr. Bristow, who was

* A nickname in her family.

THE CHURCH AND VICARAGE AT SELBORNE IN GILBERT WHITE'S TIME

[To face p. 66, Vol. I.

absent from ill-health; and resided at the vicarage, where he occasionally entertained friends.

On January 29th, 1752, Mulso writes:—

"I hear you are snug at the vicarage, where it is to be presumed that you are preparing something for the world. Sermons or satire must come from him who has left the world. . . . I long to hear some of *your* sermons in Sunbury Church, and am glad you are in strong exercise, because strengthening of the chest is I believe strengthening of the lungs. . . . You are happy in both the turn and power to labour, to journey, to harden: and if one was to address you one should begin like an epitaph 'siste viator'!"

In February, 1752, a short visit was made to Oxford, no doubt on the business of the office of Proctor to the University, which had come to Oriel for this turn, and was at Gilbert White's option. Though by no means without precedent, it was perhaps a little unusual for a non-resident to serve this office. However, his claim was recognised. He resigned his Selborne curacy in March, and on April 8th, 1752, became Junior Proctor.

Mulso, who had previously urged acceptance of the office, writes on March 28th, 1752:—

"I think you have paid the University a great Compliment in accepting of the Sleeves, for as I take your genius, you are rather Atticus than Tully. 'Otium cum Dignitate' is your motto and turn. And the green Retreats (for they begin now to be the *green* Retreats) of Selbourne afford more serious pleasure to your contemplative mind, than the 'frequentis plausus theatri' can to your Ambition. I have a longing desire to see you in your new station, but then I want to bring in each hand a girl.* . . . How prettily would they adjust your Sleeve and give a more rakish air than suits the Academic Form! How would they admire the Tuft, and how would they fancy the Flap! . . . I have had a great cold. Heck and Missy resolved to take a post-chaise and come and nurse me at Sunbury. How did they wish that Mr. Proctor White was here with them! The weather was tempestuous, but we read Pope, etc., longing for your indicative finger to point out the beauties, though it affronts our judgment by preventing it. . . ."

He goes on to mention another "scheme of pleasure" which might prevent the Oxford visit.

"And perhaps of all years this is that in which you might be least glad of our company. And I must own it would have an odd sound, when the Provost sends to know what noise that is in his college at One in the morning, to have him answered, 'Sir, it is the Proctor with two girls and the mad Parson of Sunbury.' So see you to it. . . . I reckon you often turn your eyes southward, and pine after the romantic vicarage with the 'pensile nest-like bowers' of Selbourne."

Probably upon the authority of Edmund White (Vicar of Newton Valence), Professor Bell† relates the following amusing occurrence during the Proctorship. It appears that, as the Proctor was going his nightly round, he found a dissipated undergraduate lying in a drunken sleep at the back of the Schools. Next morning the culprit was dismissed with a

* Evidently his *fiancée*, Miss Young, and his sister.

† *Vide* Bell's edition, vol. i. pp. xxxii., xxxiii.

lecture because he had retained sufficient sense of decency to fold up tidily his outer garments! True or not, the story is at least consistent with the extreme sense of order and neatness, which in an eminent degree characterised the Junior Proctor. It is also said, but the writer has been quite unable to trace the story to any authority, that the future historian of the Roman Empire, Gibbon, who was certainly in residence as an undergraduate at this time (since he matriculated at Magdalen College on April 3rd, 1752, and remained in residence at Oxford for fourteen months), was upon one occasion proctorised by the future historian of Selborne.

Residence at Oxford during the year 1752, though it necessarily caused some interruption to the regularity of the entries in the *Garden Kalendar*, was far from putting an end to them altogether. The *Garden Kalendar* for that year, like that for the preceding year, contains frequent entries of the planting of trees and shrubs in the garden and grounds of the Selborne home.

The account-book kept by Gilbert White recording the "Expence preparatory to and during my year of Proctorship in the University of Oxon., 1752," has been published in its entirety by Professor Bell,* with the true comment that it exhibits "in its simplicity that combination of genial kindliness and

* *Vide* Bell's edition of 'The Natural History and Antiquities of Selborne,' vol. ii. pp. 316–346.

generous hospitality, with habitual prudence, punctilious formality, and methodical habits which was so characteristic of his after life." To Mr. Bell's book the reader who desires to look into some phases of ordinary college life at this period may be referred. Here it will suffice to say that Mr. Proctor seems to have fully entered into the social life of the University, going to concerts on "choral nights," entertaining his friends with "A bowl of rum Punch from Horsman's," occasionally visiting the "Coffee House," and playing cards in the common room, while entries for cleaning his gun and the purchase "from Woodstock turnpike" of "a Spaniel of the Blenheim breed," show that he still continued to shoot.

Nor were his friends outside Oxford forgotten. His brother Thomas, always interested in botany, came to see "the Physic Garden"; Mr.* and Mrs. Whiston; his cousin, Baptist Isaac; his sister, Mrs. Woods; and uncle, Mr. Snooke (of Ringmer); and several of the Mulso family, to entertain whom he returned to Oxford in August for a short time. Nor were the "poor of Selborne" forgotten by his charity.

During this time he served the office of Dean of his college, the most important post after the Provostship.

On December 21st, 1752, Mulso writes:—

* Doubtless the senior partner of the firm of Whiston and White.

"You have now passed through a good deal of your Oxford confinement, yet much remains; what a pleasure will you feel when this honourable clog is taken off and you at liberty to range the country as you were wont!"

In January, 1753, when Gilbert White was at home for the Christmas vacation, his brother John, who had some mechanical talent, was undertaking, with some pecuniary assistance from other members of the family, a work which must be very familiar to all who know Selborne—the construction of the well-known zigzag path from the bottom to the top of the Hanger. This work had been apparently commenced in the preceding autumn, since Gilbert White's account-book contains the entry—

		Li	*s*	*d*
1752.	"Sep[t] 29 Gave towards my B[r] John's Ziczac up Arbour Hill	00.	01.	00."

and in the following October there is mention of a second subscription.

John Mulso, writing on January 27th, 1753, says:—

"I have great ideas, but I suppose not adequate, of your famous Zigzag, but I intend not to strain my fancy; that being the farther distant from the real greatness of the work, I may enjoy the Surprise the more, but why do I say enjoy? I do not even foresee a visit to Selbourne. I fancy I shall like the alteration of your Hill better than the alteration of your verse, for unless by the difficulty of getting through the one, you would signify the labour of climbing up the other, (which is wrong to the very design) I do not think the crassitude and impediments of the line compensated even by the descriptiveness of it. . . . Our girls are clear that the affair between you and *one Jenny* is quite serious. Missy is very fond of the thought, being much taken with the lady; but you was so grave with me in the post-chaise that I dare not add to their opinion anything but my applause of the lady. However that may be, I daresay that she is very instrumental in softening the Rigour of your Oxford confinement, and often prevents your forgetting family life. . . . Miss Heck has got a new cantata for you, and if you come before her pipe is quite stopped you may command it."

Who the lady referred to was does not certainly appear, but his friend had rallied Gilbert White upon a certain "Jenny" in Oxford some years previously. Very probably the lady in question was "Jenny Croke," with whom Gilbert White journeyed "from Selborne to Oxon. in a post-chaise," and subsequently presented with "a round China-turene, being prevented paying for ye post-chaise."* From other entries in his accounts it seems that Mrs. Croke, Jenny's mother, kept a haberdasher's shop at Oxford, and sometimes collected the rents from Grandmother White's tenants at Oxford. The frequent occurrence of a present to "Mrs. Croke's man" shows that Gilbert White often visited at her house.

At the top of the zigzag path may still be seen a rough-hewn, rudely pointed stone of moderate size, which was placed in position by the brothers White at a somewhat later date. It is a little curious that,

* *Vide* Bell's edition, 'Gilbert White's Account-book,' vol. ii. p. 323.

THE "OBELISK" AT THE TOP OF THE ZIGZAG PATH AT SELBORNE

[To face p. 72, Vol. I.

though the Naturalist is by no means forgotten in his native village, all memory of his connection with the Zigzag and this stone, which is now termed the "wishing-stone," and gravely conjectured to be of Druidic origin, is now completely lost. This, however, is certainly one of the stones which he dignified by the name of "obelisk." His account-book for 1756 contains the entry—

"Feb. 7.—Gave Lasham for opening a vista for an obelisk."*

And later, in 1758, the entry occurs—

"Oct. 14.—Bringing rock to Zigzag and making area."

As the rock, or "obelisk," is a piece of sandstone, it must have been brought from the forest side of Selborne, where there is plenty of it. These works, together with the Hermitage, towards which 5*s.* was subscribed in April, 1758, no doubt afforded great pleasure at the time to the brothers; and however unimportant in themselves, likely enough really added to the strong attachment they all formed to Selborne.

Mulso, writing to his friend in March, 1753, after congratulating him upon his prospect of recovering his liberty, continues—

"I am glad you name May for the time of your enlargement, because by that time the Prince will be at Kew, and

* The particular "obelisk" referred to in this entry was probably placed in the field at the bottom of the Hanger, opposite "The Wakes," whence it could be viewed through the "Vista." A stone of some size, which probably formed part of it, still stands in this position.

you and I will go over to dine with my uncle through some of the most delightful riding in England. . . . Sir Philip [Musgrave of Eden Hall] has carried Chardin [Musgrave] to wait upon my uncle, but I do not apprehend that he has taken any great fancy to him, for your story has given him a very great dislike to Him"—

a significant mention of Chardin Musgrave, who was shortly afterwards to become Gilbert White's successful rival in the candidature for the Provostship of Oriel.

Shortly before this time (in February) Benjamin White married Miss Ann Yalden, daughter of the Rev. Edmund Yalden, Vicar of Newton Valence. In May, Mrs. Charles White, the wife of the Rector of Bradley and Swarraton, died.

On May 2nd, 1753, Gilbert White went out of his office of Proctor, and at the end of the month set out for "London, Sunbury, and Selborne." Writing at this time, Mulso begs that "when you come you will bring your Selbourne voice that you may be heard in our church." Later on, in July, he paid a visit to his relative, Dr. White, at Christian Malford, Wilts, on his way to "a seven weeks' season at the Hot well at Bristol, from July 9th to August the 30th."

A slight feverish attack seems to have been the cause of this visit, since Mulso, after mentioning that he himself, like his friend, had been suffering from "Inward Heat," continues—

"I carried your letter to Kew, where it was read by Bishops, Priests, and Deacons. My uncle laughed heartily at your *Hectic Heat*, and my aunt said that half such a joke was a serious Proposal, and we laughed beforehand at the fright we suppose you in at reading this, by which you find you have drawn yourself into a Præmunire. . . . As to your Rags and Chips, Heck totally disdains all sinister and Canidian Use of the same; and protests she trusts to no Foreign Charms for your demolition"—

a passage which, in itself, clearly disposes of the statement, above noticed, that this lady was the object of an unrequited attachment on the part of her friend.

CHAPTER V.

ON September 9th, 1753, Gilbert White became Curate of the small rural parish of Durley, near Bishop's Waltham, Hants. In the latter place he had lodgings, paying Mr. Gibson, the Rector of Bishop's Waltham, £20 for a year's board; but his curacy probably only involved Sunday duty, since he appears to have constantly visited his family at Selborne, riding backwards and forwards on "Mouse."

It is noted by Mr. Bell* that, during the year and a half his curacy of Durley continued, his expenses exceeded the receipts from his curacy by nearly £20, from which he infers that Gilbert White was influenced by "unselfishness and true Christian liberality in accepting such a poor cure." Perhaps, however, the fact that this curacy, though a poorly paid one, did not involve a continuous severance from his home at Selborne may have influenced him in accepting it. Certainly the *Garden*

* *Vide* Bell's edition of 'The Natural History and Antiquities of Selborne,' vol. i. pp. xxxiv., xxxv.

Kalendar was continued very much as before, and in his handwriting, until the end of October. It commences again towards the end of February, 1754.

Writing to his friend on October 18th, 1753, Mulso says :—

"I am well pleased to hear from yourself that you are settled for a Time and in a Place to your liking. . . . I who came to attend your Glory, when you glistened in your Velvet, and powdered your grand wig every day, find an equal desire to wait upon the weather-beaten curate of Durley, in his dirty Boots and dripping Bob. 'Omnis Aristippum decuit Color et Status et Res.' You are the Philosopher. . . . I remember the hospitality of Gibson on our way to Gosport from Winchester. I likewise remember the romantic scene which you mention in your letter, which paid me for being more than half starved in going to see it. . . . My uncle [the Bishop] asked me, how you came to take a curacy? I told him because I knew that it was your sentiment that a Clergyman should not be idle and unemployed"—

a remark of some importance in view of certain strictures to which it will be necessary to refer presently. It is to be noted that upon this sentiment Gilbert White acted during his whole life.

In October, 1753, the Curate of Durley posted to Oxford in order to resign the Deanship of Oriel. Returning to Durley he bought, and no doubt carefully studied, a new edition of 'Miller's Gardening Dictionary' and 'Raii Methodus,' the latter a book to which he constantly refers in after years.

On December 24th, 1753, Mulso writes :—

"You will find enclosed with this the song you desired me to procure you. . . . I see your good Father setting himself to his Harpsichord and trying it above forty times; for unless his hand is more in than when I saw him, he will be some time recovering so full a Tune. Harry turns up his honest face into the air, and pours out his part presently. Harry has a good ear."

The family was a musical one. The diaries kept by Harry White, when Rector of Fyfield, continually mention musical evenings at his house, which contained, besides a piano, a harpsichord, spinet, violin, and violoncello. They show, moreover, that he could not only play but tune a harpsichord.

Cut off as he seems to have been from his home during the winter months, the Curate of Durley found time to indulge in a little shooting. He also entertained "the Waltham gentry," and in March went "with Ladies to a puppet shew," presumably at some fair. With the opening of spring the Selborne garden received much of his attention, the *Garden Kalendar* containing numerous entries of improvements, alterations, planting, etc., as the year went on. His headquarters, however, were still at Waltham—unless indeed the Selborne house should really receive this appellation.

However, other matters than gardening received attention, since Mulso writes on February 9th, 1754 :—

"I presume by your enquiries after modern copies of Diogenes and Aristippus, that you have imitated the 17th Ep. of the 1st book of Horace. If so, I look upon it as lawful Prize, and put in for a Sight of it. . . . It is an odd time of the year for you to see gardens in; I had rather see them in your description than in reality. You have carried me round a very pleasant Tour."

On March 26th, 1754, Mulso wrote again :—

"Am I to suppose that your life has in it a great deal of sameness or a great variety, that you are so bad a correspondent? . . . I was in town about three weeks ago and there saw your brother Jack, and a very smart parson he makes. He knew nothing of you and your motions. Meanwhile you are like a comet, who in your 'secreto itinere et certo errore' are sucking in little worlds of knowledge and funds of light, with which when you roll this way, you are to astonish and eclipse us with your blaze."

He goes on to request his friend to find him a horse, and proceeds to describe perfection in horseflesh, from his own point of view, which he expects to procure for twelve pounds at the most, "which I hope in Hampshire is a tolerable price."

At this time an attack of illness suffered by Gilbert White's father at Selborne drew the following encomium from Mulso, who wrote on April 5th, 1754 :—

"I hope you are now on the safe side, and that Mr. White is out of all danger. Knowing Mr. White as I do, I can testify the truth of those virtues, which you, with all the warmth of filial piety ascribe to him. No good man can die without being a great loss, and I know in how many respects your excellent father would be so. Amiable family! where the parent is loved for the children's sake, and the children for the father's."

He goes on to accept the loan of "your little horse," which, however, was promptly returned. Writing on May 7th, 1754, three reasons for this are given:—

"First, he is broken-winded, and wheezes so loud that my heartache will do me more harm than the air good; next, I was forced to carry his head; then he is intolerably shabby, and will not go on a handgallop without constant incitement of spur and whip. Nor can I conceive, what with age and infirmities, how it is possible for him to keep company with a horse fourteen feet [*sic*] high. I like his paces pretty well, and believe when he was young, that he was an agreeable creature, but would you have the Prince of Wales know me by such a horse, as he did by my lame one?"

Notwithstanding the horse episode, the friendship continued unbroken, the Curate of Durley being invited to visit Sunbury. His friend, writing on May 28th, 1754, promised

"to find a method to trot about with you while you are in this part of the world, either by borrowing or hiring. You shall compare accounts with the travelled Mr. Aldrich, and by your descriptions of your native Selbourne you shall

'Shame Vallombrosa and her Tuscan glooms.'

. . . I hope your uncle White is well. I suppose he comes often over to Selbourne, for I suppose you seldomer visit Bradley than Selbourne now."

The visit was paid in June. On September 16th, 1754, Mulso writes:—

"Do you go to Oxford upon the scheme you mentioned? Is there a hope of its succeeding? or are you to sit down at Waltham this winter, and warm yourself by the widow's fire?"

A remark which probably has some connection with the following entry in G. White's account-book:—

"Oct. 14.—Postchaises from Bradley to Oxon. on Mr. Whiting's death."

The letter proceeds to mention more improvements in the Selborne fields, where "six gates one above another in Perspective" had been contrived, apparently between "The Wakes" and the Hanger. A little later, on November 13th, 1754, some accident to his friend's knee, which nearly had serious consequences, called forth these comments from Mulso:—

"I don't know anybody who would have felt the severity of being a cripple more livelily than yourself, for if you once come to confinement you are gone."

Probably this accident detained the patient at Oxford, where he stayed till the middle of November; whence he posted "the day after Harry's election to Bishop Robinson's Exhibition" (at Oriel), which no doubt was the object of the journey to Oxford.

The letter continues—

"I had not heard the circumstances of Whiting's death. I heartily wish that all party rage had died with him; and that not so much because we have increasing obligations to the present Family, but for a real regard for the University, which is in a very low consideration for the sake of a parcel of Fools, who are a disgrace to it in every view, and are of the most contracted hearts of any set of men that I know. I am very glad that Dr. Bentham* is to have so long a grace, but do not see how it naturally comes to pass, and if by an interest it is a good sign I hope that he is stronger in it than I once thought. . . . How did Whiting salve to his conscience the holding his fellowship six years in wedlock? Was this one of his indefeasable rights?"

These remarks are of importance as showing the lax manner in which, at this time, the college statutes respecting the avoidance of fellowships were carried out. It will be seen that a question of this kind was thought to have arisen a few years afterwards in the case of Gilbert White, when the Provost of the time (Dr. Musgrave) acted in a very different spirit.

On December 24th, 1754, Mulso writes again:—

"I indulged myself lately at Mr. Hubert's, who winters with us, in talking of you, and I had no occasion to exert my rhetoric to gain you favour. You was spoken of in a very handsome manner by Mr. Hubert's family, and Phil desired to sum up the character by saying, in short, he is *the Gentleman.* I don't know whether you, who was always averse to a distinction of Place's being made a distinction of Merit, will be satisfied with this corollary; but I can assure you it was, in the sense of him who used it, as perfect and compleat a one as any of Euclid's. . . . I think I have wished you joy of your brother Hal's success. You have established your character I presume of a Plotter. I am glad to hear of Dr. Bentham's long grace."

* Fellow and tutor of Oriel College, who had recently accepted a living and married.

In January, 1755, the oldest (in every sense of the word) inmate of "The Wakes"—Gilbert White's grandmother, widow of the Vicar of Selborne—died. She had attained the great age of ninety-one, as her grandson was careful to record in the burial register.

On February 10th, 1755, Mulso, who had heard a false report of the death of Provost Hodges, writes:—

"As to your Oriel news, it affords a Prospect, but I suppose you hardly entertain solid hopes of any present preferment by it, tho' it may make way for Harry. Young Mr. Shaw of Cheshunt would yesterday have persuaded me that Dr. Hodges was dead and that you was going to be Provost in his Room. I should have liked one part of the news very well if true, but I know you would have invited me to condole with you upon the death of the worthy Provost."

These, however, were dreams of the future. After the usual visit to Oxford at Easter, Gilbert White made arrangements to take up temporarily another curacy, that of West Dene, or Deane, in Wiltshire, not far from Salisbury, and just on the borders of Hampshire. Writing to him there on April 8th, 1755, after mentioning that his uncle was looking towards Salisbury for a place to put his mitre in, Mulso continues:—

"Perhaps it is better being a Bishop than a curate there, but indeed I am not sure, for as a friend of mine* sang,

'About content why keep we such a Riot?
'Tis here, at Cowbitt,† if we could be quiet.'

* Evidently the friend he was addressing.
† In the Lincolnshire fens.

You give me pleasure in hearing of the stand made against the perverse party at Oriel. I would the Provost should live till you succeed him (if that is English; it sounds Irish); and then if I have a son, he goes to Oriel."

Before going to Dene, however, he found it necessary to go on May 1st to have another "7 weeks season" at the Hot Well at Bristol, returning to Dene; where he took up his residence, at the beginning of July, as curate to Mr. Edmund Yalden, Vicar of Newton Valence and of West Dene.

On May 28th, 1755, Mulso, who had been promised a canonry by his uncle, writes again :—

"The everlasting prebendaries of Peterborough are all in good health as I am informed by the Bishop. . . . I am charmed with your description of the site of your parsonage house. . . . You threaten me if I do not send you word of some preferment falling to me, that you will supplant me by making love to my cozens: truly (but this is entirely *entre nous*) that Sap is begun, or I am much mistaken; so that, if it does not fall soon, I may chance to hear that though a nephew is dear, a child is dearer; which on such an occasion would be a melancholy Truth which I could not gainsay. . . . I travelled very pleasantly with you to the Hot Wells in imagination, yet your first day's journey of 34 miles would in reality I believe have finished me. . . . I sincerely wish you had a living like Dene, and the thorough good sort of damoiselle that you mention, that your wishes might be compleated, and that I might say, I know and have a value for Mr. White of such a place, for now you are of no place for a fortnight a perfect Cup of unsettled."

THE HOT WELLS AT BRISTOL IN THE EIGHTEENTH CENTURY

[To face p. 84, Vol. I.

That his new abode was not entirely to Gilbert White's liking appears from the following sentence in a letter of September 18th, 1755, from Mulso:—

"I hope this will find you well and reconciled to your situation; which, tho' you have as much true Philosophy as any man I know, yet is not to your taste, if it is really solitary."

He managed, however, to amuse himself sometimes, for he visited "Sarum, Stonehenge, and Wilton [House]" with his relatives the Barkers, and paid "a man for showing me troufle-hunting," of which he afterwards wrote in his book. Sometimes, too, he went over to Selborne, and continued to write his *Garden Kalendar* there; riding over, of course, which accounts for the not infrequent purchases of "black leather riding-breeches." In these solitary excursions on horseback he had good opportunities, which were certainly used, for observing natural objects of many kinds, both flora and fauna. His friend, who had now been engaged to be married for ten years, continued:—

"Do not you fix your eye upon Cromhall or Tortworth, [Oriel livings] or indeed upon anything particular; for the fixt eye will be an aking one, believe me. I have looked at Peterboro' til it now seems lost in a mist."

In the autumn of the year 1755 Gilbert White undertook duty for Mr. Yalden at Newton Valence, close to Selborne, as well as at Dene. For some considerable time his father's health had been giving anxiety to his family; and this, of course, made his son all the more anxious to return, probably to the vicarage at Selborne, which place, as it happened, was from this time to be always his home.

Mulso writes on December 6th, 1755:—

"I am sorry that your duty is so increased as to be grown troublesome, yet methinks I am glad that you are got near enough to be more a comfort to Mr. White. . . . Tell me of your Vases and Obelisks; let me see them in imagination, if not in reality; I have a pretty good idea of your grounds, place me at my proper distance, and let me see your Antonines and Trajans with their Egyptian Hieroglyphics. . . . Cannot you make that idle rogue Jack* change duties for some time, when he is disposed to visit his father."

Early in 1756 his final visit, for the time, to Dene, was apparently made; and from this time the *Garden Kalendar* gives evidence of leisure spent in the Selborne gardens in planting many sorts of shrubs and plants, some of them received from brother Thomas in London, "in the basons in the field," "opening a vista for an obelisk," etc.

Writing on April 23rd, 1756, Mulso says:—

"Now you begin to see the effects of your vases and obelisk amongst the green Hedges. Your gates still remain mysterious, but your very exact and strong description has set your other improvements before my eyes."

He continues—

"Pray, Gil, let me know a Truth. You stand indited by the name of Gilbert White Clerk, for that you having the

* John White, Gilbert's brother, was at this time a curate in London.

whole and sole Property of a thing called a Sermon wrote by Miss M——o, keeping it from the family of the said Miss out of a pretended Pride of having a manuscript value 10,000 &c &c &c: have yet let this manuscript escape out of your Possession; Mr. Proctor Turley of Sunbury having proffered me to get sight of the same; as we suppose, tho' not yet proved, by means of his uncle Brown, Bookseller, who is acquainted with Mr. Whiston, Bookseller, & B. White ditto, brother of the said G. White ye Delinquent: who is mainly suspected of having made undue communications of these Lady-Favours, a thing unpardonable, and till this time unsuspected in the said G. White. Please to clear up these affairs, before condemnation is passed in the King's Square Court."*

The above passage is scarcely what a brother would have written to a friend who cherished an unrequited passion for his sister. That friend was at this time constantly engaged in embellishing the Selborne grounds.

From the *Garden Kalendar* :—

"May 14th. Set up my first oil-jar vase at the bottom of the ewel close, with a pannel only in front; mount pedestal and vase nine feet high.

"19th. Set up my second oil-jar vase at the top of the broad walk, with a face to the cross-walk; mount pedestal and jar some inches above nine feet high."

In opening out his father's grounds and making ha-ha's and cutting vistas, Gilbert White was following the fashion first set by William Kent, painter and landscape gardener (1685–1748), who did more than anyone to abolish walled, closed-in gardens,

* The residence of Mr. Mulso, senior.

with monstrosities in the shape of clipped trees and hedges.

In June he was able to spend a holiday of about a month at Sunbury with Mulso, who, after a long engagement, had just been married to Miss Young, receiving on the occasion a soup tureen, which he thought "a great beauty," from his old friend—a fact which confutes the statement sometimes made that wedding presents are of comparatively recent origin.

On August 23rd, 1756, his friend again writes :—

> "You will remember that you have two commissions for me; one to get me a horse for ten guineas, and the other to get me a curate. I had rather have both from your Recommendation than from any other man's in England, because you are more able to see with your own Eyes than any man I know besides. . . .
>
> ". . . How goes on the Pyramid? Have you clapped on the Handles to your urns? Does your father like your Improvements? They must make a sort of new scene to him at home."

Purchases of a setting-net and powder and shot, at this time, show that the Naturalist's sporting days were not yet over. In October a week was spent at Oxford, where he paid "for Divinity Disputations £3 3. 0."; returning home by way of Sunbury, whither he went to "sacrifice your Pleasure to my convenience, as usual," as his friend, who was unwell, wrote. These visits to Oxford, in addition to the Easter one, which he never missed, are by no means uncommon; a circumstance which is scarcely consistent with a statement which has been made in recent years, but without any sort of evidence to support it, that Gilbert White's relations with the members of his college were not entirely cordial. Mr. Jesse states,* "on the authority of one of his nephews," in a short biography of Gilbert White:—

"As long as his health allowed him he always attended the annual election of Fellows† of Oriel College, where the gentlemen Commoners were allowed the use of the Common-room after dinner. This liberty they seldom availed themselves of, except on the occasion of Mr. White's visits; for such was his happy, and, indeed inimitable manner of relating an anecdote, and telling a story, that the room always filled when he was there."

Perhaps the following extract from a letter written in 1783 by Gilbert White's niece, Mary, to her brother, Thomas Holt-White, at Oriel, may refer to one of these stories:—

"I want to know when you got into your rooms and how you liked them, and Oxford in general, but that I suppose you will say at present you can be no judge of. My Uncle White says he remembers calling on two brothers who inhabited the same rooms, and desiring to borrow a pen and ink, and his being rather surprised with the answer that they really never had possessed any since they came to College, tho' they had been there two or three years. I hope this fashion is not general, indeed he mentions it as *rather* a particular thing."

* Bohn's edition, 1854, p. xi.

† For this we should read "The annual audit of accounts."

On November 6th, 1756, Mr. Yalden settled with his curate for supplying the curacies of Dene and Newton "for a year and a quarter" to that date.

The curacy of Selborne was, apparently, for the second time undertaken by Gilbert White at this time, since he records the payment of £15 from Dr. Bristowe, the Vicar, on this account in April, 1757; and a further sum of £10 "in full" on July 4th of the same year.

In November, 1756, the following entry appears:—

	£	s.	d.
Year's Board for self and Mouse [his horse] to my father to Nov. 5	40	00	00,"

the vicarage house of Selborne having merely afforded sleeping accommodation.

On January 13th, 1757, Mulso, who had received no news of his friend for some time, writes:—

"I conclude that the mornings were spent in riding, and the afternoons in Burnet's 'History of the Reformation'; and I had two pictures of you before my eyes; one was the scene of Dr. Bristow's parlour, and the other your own. In the first you was a solitary Figure, and as hard at it as Duns Scotus; in the other I had represented the very pleasing Figure of Miss White, with some Housewifery in her Hand, and giving now and then an agreeable interruption to your Labours. Harry was cast into the Shade, and it was doubtful whether he was studying or no; but this confusedness of my ideas was owing to his being so perverse as not to confirm his Image in my Imagination when I was in Hampshire."

He goes on—

"I am glad that you got to some success at last at Oriel, and I wish you joy of tiles and thatch. You, who can make £20 go farther than I can £40, have a pretty little increase by this curacy. . . . I am glad you are so well at the Grange, and that my Lord* is so open with you. If Expectation makes the blessing dear, you and I are like to have the dearest blessings in the World."

This preferment was the perpetual curacy of Moreton Pinkney, in Northamptonshire, about to become vacant through its holder's promotion. The matter was not, however, finally settled till December, 1757.

Meanwhile an event of importance occurred. The Provost of Oriel, Dr. Walter Hodges, died on January 14th, 1757; and Gilbert White became a candidate for the office, but unsuccessfully, since on January 27th Mr. Chardin Musgrave, fourth son of Sir Christopher Musgrave, of Eden Hall, was elected.

On February 24th, 1757, Mulso writes:—

"I received yours from Oxford Jan. 28th with an account of your election at Oriel; but the first account that I received of it was from Sir Philip Musgrave, who had a variety of fortune happened in his Family in the space of a Fortnight; for he had a son and Heir born, his sister Spragg left a widow with about £10,000, and Chardin chosen Provost: and since that time died Mr. Beckford from whom

* The Lord Keeper. At this time Sir Robert Henley, a Hampshire neighbour and friend of the White family, who afterwards became Lord Chancellor.

my Lady had I believe expectations of getting something. Sir Philip spoke of Frewen's proceedings as not very handsome upon the occasion, and as putting them to the necessity of applying out of the College, when they would have had it determined there. How that may be I do not know; but as you have not been the man on this Preferment, I am not sorry for the success of Chardin; unless it may prove any obstruction to your designs for your brother Harry, which I greatly hope it will not. If Chardin behaves in his post with the good sense and judgment that Sir Philip talked of it, the College will have no reason to repent their choice. I have still a good deal to say to you on this subject, but intend to say it and not to write it."

After referring to the late Provost's legacy to the College, he continues—

"I am sorry you mention no remembrances to you or your Family, but your brother Harry has the greatest loss in this Friend."

How far his non-success was a disappointment to Gilbert White can now only be a matter of conjecture. His friend, however, regretted it, and wrote again a little later, on March 19th, 1757 :—

"With regard to the affair of Oriel, I heartily wish you had put yourself up from the beginning, if anything we could have done would have given you success. But yet I think you judge of the issue of Chardin's election otherwise than you will find it turn out: at least I hope so. But I believe this subject will do better to talk upon than to write upon; for tho' we agree in opinion pretty well, a little error of the Pen may make a Discordancy, and I find that with you Oriel men goes farther than I think, in justice, it should. As to your Brother, nothing in my talk with Sir Philip had any reference to Him; but to the impartiality that he had advised him to observe, as the best method of serving the College: upon which principle, if the Provost observes it, your Brother may be successful."

In obtaining a Fellowship, no doubt, was the writer's meaning.

In April, 1757, Gilbert White visited Oxford as usual, and a little later (on May 12th, 1757) his constant correspondent writes:—

"Curate or not curate, I find you will travel, and a restless animal you still will be till I find you squatted down in Fat-goose living."

Though he does not seem to have displayed any very great anxiety for a living at any time, his friend recurs to the subject on July 14th, 1757—

"I looked, my dear friend, to have you nearer the Town before now, and besieging the Portals of the Lord Keeper; I have great expectations from that quarter for you, and his preferment has given me particular pleasure from knowing him to be a near acquaintance and friend to you and yours. Sure there are more ways than one of vacating a Fellowship!"

From September 4th to October 23rd, 1757, Gilbert White acted as Curate of Dene and Newton in the place of his cousin, Basil Cane, who was in Wales. The living of Moreton Pinkney became definitely vacant in October, 1757, when its incumbent, Mr. Frewen, was presented by the college to the living of S. Mary's, Oxford. On October 17th the Provost's memorandum-book contains a statement of this presentation, and continues, "by which Moreton Pinkney will become vacant; agreed to give it to the Senior Fellow, who will serve it in Person." With regard to the latter part of this entry it should be especially noticed that residence had not previously been required of any Fellow who accepted this preferment, which was so small as not to vacate a Fellowship. The next entry referring to this subject is—

> "Dec. 15. Moreton Pinkney given to Mr. White as Senior Petitioner, tho' without his intentions of serving it, not choosing to waive his claim, tho' Mr. Land would have accepted it upon the other more agreeable terms to the society; I agreed to this to avoid any possibility of a misconception of partiality, as I am convinced the major part of those present did also; but agreed for the future that in any of the Tenable preferments Preference shall be given to any Fellow who will undertake to serve the cure, before a Senior who would put in a Deputy."

It is noticeable that the sudden access of virtue which inspired the resolution to insist upon the residence of the new incumbent was coincident with the appointment of a Provost, who had not only recently been in rivalry with Gilbert White, but, as plainly appears from a former letter of Mulso's, had also for some time been regarded by him with some amount of dislike, which was probably, and not unnaturally, reciprocated.

On his part, Gilbert White no doubt urged, as was the fact, that in the case of this very small cure residence had never hitherto been considered a condition of appointment, and that the new stipulation ought not to set aside his claim to hold it upon the customary conditions. Though he occasionally visited, he never resided at Moreton Pinkney; which continued to be served by a curate, whose stipend left some £30 to the holder of the living, who was formally licensed to it in London, by the Bishop of Peterborough, in May, 1758.

On June 10th, 1758, Mulso writes:—

"Methinks I see you very busy at your Father's map of France, tracing out the environs of St. Malo and Brest. We have very favourable accounts at present from all quarters. . . . The year is as beautiful as ever I saw. I long to be in your Grotto [The Hermitage]—*gelidis sub montibus Hæmi* is your situation."

In July, 1758, the Vicar of Selborne, Dr. Bristowe, who had been for some time incapacitated, died. Mulso, after condolences upon the death of such an old friend, remarks, on July 13th, 1758:—

"I suppose you have the care of the church upon you till the successor arrives. I beg of you to contrive to get a great estate, to be enabled to live on at Selbourne, to be the friend of the Poor, who have now lost one, and may in a few years lose another [*i.e.* Gilbert White's father]; and prevent that sweet place from decaying into the very Den of Poverty and misery: capable as it certainly is of the highest improvements, and of being one of the most enchanting spots in England."

During the summer the improvements at Selborne continued. The zigzag path received a cleaning, "rocks" were fixed at the Hermitage, and a little later there is another entry of "bringing rock to the Zigzag and making an area," and two wickets were added to the Zigzag. These "rocks" about the Hermitage have, like the building, long disappeared from their position; but they are clearly seen in the vignette view of the Hermitage, by Grimm, which appeared in the early editions of 'The Natural History and Antiquities of Selborne.'

Writing on August 28th, 1758, Mulso says:—

"You said you wished (without prejudice to anybody) you was set down upon the living of Deane; I wish you was with all my heart, for ever since you mentioned the Isle of Wight, I have been sea-sick at the thoughts of crossing the water. And yet I know it would be impossible for me not to come to see how you would improve any place that you was settled in: but pray, if you can, let Fat-goose living be upon Terra firma. . . . Mrs. Mulso longs to see your alterations at Selbourne, neither has she less curiosity to know your ingenious brother.* Every hot day she wishes to be transported to the peaceful Hermitage."

The *Garden Kalendar* records on September 5th—

"Eat a brace of Cantaleupes at the Hermitage."

So pleased were the contrivers with their Hermitage that the elder brother broke out into verse.

On September 12th, 1758, Mulso writes:—

"You give us hopes that we may see Miss White as well as yourself at Sunbury. You seem to be confined by the sequestration [of Selborne living], and yet this detached little piece

* Harry, who used to appear as the hermit.

of news seems to mean that we shall see you before you are released from that confinement. . . .

"I have been looking over that copy of verses which I have in my possession under the title of the *Invitation*, in order to insert the beautiful Lines which you have added in honour of your Hermitage,* and to alter the two rhimes which I thought abominable, but I can find no such lines as begin with 'Oft on some evening,' etc., and I believe that you sent some to Miss Mulso that are different from what I have by me, which came out of the Fens of Cambridgeshire."

The letter continues—

"Our Swallows and Bank Martins have long been dashing into the heights, as if they had called their convocation to debate upon departing. You know you was struck with this sight last year."

Swallows and their doings always had a peculiar interest for the Naturalist, who recorded a little later, on November 2nd, 1758, in the *Garden Kalendar*—

"Saw a very unusual sight; a large flock of House-Martens playing about between our fields, and the Hanger. I never saw any of the Swallow-kind later than the old 10 of October. The Hanger being quite naked of leaves made the sight the more extraordinary."

* Evidently the lines—

"Or where the Hermit hangs the straw-clad cell,
Emerging gently from the leafy dell," etc.
Vide "The Invitation to Selborne."

CHAPTER VI.

On September 29th, 1758, John White died. The news of his father's death was at once communicated by Gilbert White to his friend Mulso, who wrote in a letter of condolence on October 5th :—

"With regard to his family, who will always remember him with the tenderness and honour due to his memory, he was spared to them till they were all grown up to such a state as to be at no loss for a method and settlement in Life; and what little is wanting to Harry would have naturally fallen to your care, had he lived longer, from his retired way of life."

He goes on to remind his friend that "solid is the blessing of having had a good parent." This blessing, there is no doubt, the Selborne family enjoyed. The father of a family of sons, none of whom were quite commonplace men, John White himself does not seem to have exhibited any great strength of character; yet he earned, and no doubt deserved, his children's affections and the respect of all his neighbours. Though at his marriage he retired from practice at the Bar, he retained some connection with the Law, since he was a magistrate for Hampshire.

It may be mentioned that he directed by will that he should be buried "with as little show and expence as may be and without any monument, not desiring to have my name recorded save in the Book of Life." This injunction was observed by his children, but in the year 1811 his descendants put up a tablet to his memory in the chancel of the Church.*

The news of Gilbert White's father's death soon reached Oxford. The Provost of Oriel's memorandum-book contains the following entry :—

"Nov. 1st 1758. At a meeting I produced Dr. Blake's resignation (his year of grace being out) and pronounced his place to be vacant. At the same time hinted to Mr. White's friends that I was ignorant what his circumstances really were, but suppose his Estate incompatible, and begged he might be informed that if a year of grace was not applied for in the regular time, etc., it could not be granted."

Evidently John Mulso's ideas upon this subject were the same as Dr. Musgrave's, for at the end of this month (November 28th) he writes :—

DEAR GIL,—Though I have talked with your brother Benjn and Mr Cane, I can form at present no judgment upon what Plea you can keep your Fellowship with your Estate, so that I cannot give advice of any value to your present purposes. I cannot but conclude from my knowledge of you, that the reasons must appear very strong to you, and that you could not be tempted by Interest to do anything contrary to the Statutes of the University, or of your particular Society; and not only so, but that you can never

* On this the year of his death is wrongly given as 1759.

forget that Fellowships are a sort of Temporary Establishments for men of good Learning and small Fortunes, till their merits or some fortunate Turn* pushes them into the world, and enables them to relinquish to men under the same predicament. I am in no doubt about what you owe to the present Society in this respect; I speak only as to the general intent of your Founders and Benefactors, and as to what you owe to yourself; in which views I daresay you would be cautious of appealing to a Visitor, unless the affair was absolutely clear and creditable on your side. For visitatorial Decrees being, though statutable, something tyrannical, must make the Person appealing ill looked upon by his brethren, unless the case turns out quite fair and clear on his side. But what I now say is a caution, which is unnecessary, because I have had a long experience of the candour and honesty of your disposition, and can make nothing against the reasons which you must have, and which are quite unknown to me. I have a good deal of curiosity to see your Letter to the Society. You certainly at present owe them no Compliments to your own disadvantage. Can you not, now that one of your opposers is gone, make a push for Harry at the next Election?

That the writer's mind had been to some extent disabused of his belief in his friend's wealth is shown by this sentence in the same letter:—

"The season is chilly and dull, the chearful leaf is gone, and the poor ravage even the bough:—But alas why do I speak of this only? with you, my dear Friend, the good Tree is fallen that sheltered so many under his hospitable arms. *Non deficit alter* indeed; but if he was a little more *aureus* your poor neighbourhood would perhaps never miss their old Patron, so riveted is your natural partiality to

* *e.g.* a college living, which Gilbert White never accepted.

Selbourne. . . . I hope your brother Thomas recovers apace, and that your Fire Side is well. I pray God not to take away, but to increase your Friends and your means."

Mr. Bell seems to have seen a letter from the Provost to Gilbert White, since he remarks*—

"A letter from Dr. Musgrave, then Provost of Oriel, to Gilbert White, dated Dec. 24, 1758, clearly intimates that some representations had been made to him that Gilbert White 'had succeeded on his father's death to a very large estate,' and that on this account his retention of his fellowship, and consequently, his presentation to the College living, were inconsistent with his present position. The Provost, however, declines in the most positive manner to listen to these misrepresentations, for such they undoubtedly were, and probably made from interested motives."

This letter from the Provost would seem to have been written in reply to one from Gilbert White, in which he had explained his real circumstances, and to have expressed Dr. Musgrave's acceptance of his statement that the report of his wealth was unfounded. Writing on February 4th, 1759, Mulso says:—

"I met the Provost in St. James' Park lately, and just cursorily asked him what he intended to do with you, and he said it was in your own breast to keep or leave your fellowship, for nobody meant to turn you out if you did not choose it yourself; so I suppose that affair is settled."

* *Vide* Bell's edition of "The Natural History and Antiquities of Selborne," vol. i. p. xxxviii.

Again, on March 23rd, 1759, he wrote :—

"I am very glad that you see my solicitude about the propriety of your holding your fortune with your fellowship in the light that I could have wished. It was owing to my jealousy of your honour, not my suspicion of it, and I was satisfied that you could supply me with an answer to those who might ask me about it, though I could only give one by guess before to the same purpose. As you are satisfied of the legality of holding it, I think you are quite in the right to hold it."

When John White, a gentleman residing in the country on his own means, died, it was quite natural and proper that the college authorities should have called upon his eldest son to justify his retention of a Fellowship; but it seems scarcely so natural or proper that the following remarks respecting Gilbert White should in recent years have emanated from his own College :—

"Gilbert White, of Selborne, among the fellows of Oriel at this period has left the most lasting name, yet his College history is in curious contrast to the reputation which is popularly attached to him. Instead of being, as is often supposed, the model clergyman residing in his cure, and interested in all the concerns of the parish in which his duty lay, he was, from a College point of view, a rich, sinecure, pluralist non-resident. He held his fellowship for fifty years, 1743–1793, during which period he was out of residence, except for the year 1752–3, when the Proctorship fell to the College turn, and he came up to take it. In 1757 he similarly asserted his right to take and hold with his fellowship the small college living of Moreton Pinkney, Northants, with the avowed intention of not residing. Even at that time the conscience of the College was shocked at this proposal and the claim only reluctantly admitted. White continued to enjoy the emoluments of his Fellowship and of his College living while he resided on his patrimonial estate at Selborne, and, although it was much doubted whether his fortune did not exceed the amount allowed by the Statutes, he acted on the maxim that anything can be held by a man who can hold his tongue, and he continued to enjoy his Fellowship and his living till his death." *

That, in "taking and holding" with his Fellowship the small college living of Moreton Pinkney, Gilbert White only did what the Fellows before him, and many afterwards did without question, is certain; and, as has been already stated, the College had never previously sought to attach the condition of residence. His conduct in this matter does not appear, then, to have been exceptional or blameable. But the gravamen of the attack on his reputation—for it can be called nothing else—is undoubtedly the charge that he, by concealing his real pecuniary position from his College, continued to hold improperly his Fellowship while living "on his patrimonial estate at Selborne"—a statement, it should be observed, which is wholly based on the one entry of November 1st, 1758, in the Provost of Oriel's memorandum-book.

So serious an imputation upon his honour must be fully examined, even at the risk of tediousness, by a biographer.

* *Vide* "The Colleges of Oxford," edited by Andrew Clark, 1891, p. 121.

The fact is that the whole original supposition of the Provost, like the first impression of John Mulso, was founded on total ignorance and misconception of the real pecuniary position of John White and his eldest son. Far from being a rich man residing on a comfortable estate, the former was, at all events until the death of his wife's mother, Mrs. Holt, in 1753 or early in 1754, a man of but very moderate means with a large family to maintain and educate. At an early date in his married life he had joined with his mother in keeping house in a small residence (The Wakes) at Selborne, in which parish he at least most assuredly owned no "patrimonial estate." Indeed, it is not certain that he ever owned an acre or a cottage there, though his father undoubtedly did. At his mother's death, in 1755, he inherited one-fifth of her property, which consisted chiefly of five houses at Oxford. His will contains a general bequest of real and personal estate to his younger children. To his eldest son, Gilbert, however, he left only £20 as an executor, this son "being heir-general to the family and provided for under my marriage settlement."

Gilbert White, Vicar of Selborne, in his will only left £100 to his son John, "having already given him his full share of my estate in his education and his marriage settlement." This probably alludes to the gift of some property to his son on his marriage, since an examination of the original deed of settlement executed on the marriage of John White with

Miss Anne Holt shows that no property at all was thereby settled by the former.

The property which Gilbert White, the naturalist, now (very partially, as will appear) inherited, came, therefore, through his mother, who, subject to her mother's (Mrs. Holt's) life interest therein, brought into settlement

1. A house and lands at Rogate, Surrey.

2. Meadow land at Harting, Sussex, mortgaged for £1,000.

3. Other houses and lands at Harting.

The above three properties were (besides the mortgage for £1,000 secured on the second) subject to charges amounting to £530, and an annuity for life of £5.

4. A messuage, lands, etc., containing sixty acres in Harting.

5. A house and land at Hawkley, Hants.*

The fourth of these properties was settled (as pin-money) on Miss Holt and her devisees or appointees, and in default of any devise or appointment, on her heirs in fee-simple. Not many years after her marriage she joined with her husband and mother in selling this property, called the Nyewood (or New Wood) fields, to her relation, the Earl of Tankerville,

* It is impossible now to say what all these farms were let for, but the rent produced by agricultural land in the eighteenth century was, until the long war with France, very small indeed. One of them, "Woodhouse Farm," at Harting, was let for £34 at this time. This seems to have been the only farm which became Gilbert White's property, the others being sold to pay the younger children's portions.

whose estate of Up Park it seems to have joined. The Trustees of the settlement seem to have bought another property with the purchase money. This Mrs. John White dealt with by her will, as to which Gilbert White consulted one "Counsellor Wright," who wrote from Oxford on March 12th, 1759:—

"If you allow of the will, and pay the money charged on the estate there will be no occasion to prove it anywhere. When you pay the money it will be proper to take a conveyance of the estate from the heir of the surviving Trustee, and the younger children must join in it, and acknowledge the receipt of the money given to them respectively by your mother's will."

The four remaining properties were strictly entailed upon the eldest son of the marriage, subject, however, to a charge of £1,500 if there should be only one or two children, and of £2,000 if (as was, of course, the case) three or more younger children, for their portions. Power was reserved to the husband and wife to revoke the trusts of, and resettle by deed the Hawkley and Rogate lands (properties Nos. 1 and 5). This power they exercised in 1730, when these properties were resettled by them on their younger children.

Under the wills of his grandmothers, Mrs. White and Mrs. Holt, Gilbert White received no benefit whatever.

The reader is now generally in a position to understand how utterly different Gilbert White's real pecuniary position was from that which it was, perhaps not unnaturally, originally supposed to be by the Provost of Oriel; and incidentally he may have learned the danger of formulating a serious charge against the reputation of an estimable man, upon the slender evidence of one (confessedly uninformed) entry of a long bygone date.

If any further proof were needed that Gilbert White was really and truly not in possession of the amount of wealth considered by the statutes to be incompatible with the retention of a Fellowship,* it can be surely found in his manner of life at Selborne; where his establishment consisted during all his life of one maidservant and one man, who was gardener, groom, and footman, with the occasional addition of a labourer or of a "weeding woman" in the garden, and of a temporary maidservant, when visitors, who brought no maid with them, were entertained—an establishment which, though it satisfied the modest needs of its master, can hardly be considered that of a "rich" man residing on his "patrimonial estate."

Perhaps, however, the word "pluralist" would have most astonished the Naturalist's friends and contemporaries (and in criticising a person's conduct it is only fair to have regard to the ethics of the age he lived in). What, for instance, would have

* The provisions of the Oriel College statutes are given further on; when, in 1780, the pecuniary position of Gilbert White having become somewhat improved, it seemed necessary to again revert to the matter.

been thought of this epithet, as applied to Gilbert White, by his friend John Mulso; who, when he had obtained a living and a major and minor canonry from his uncle, the Bishop, regretfully wrote to his old friend that he could plainly see the Bishop thought his cup was now full?*

Mulso expressed his opinion of his friend's clerical modesty when he told him, after one of the numerous refusals of a comfortable college living, that he would go down to history as the man who "refused livings and served curacies." Nor can Gilbert White be fairly called a "sinecure" clergyman; since, though he paid a deputy to do his work in his little cure of Moreton Pinkney, he continued during his whole life after ordination in the exercise of clerical duty. It is true the work of a country clergyman was not then, nor is it now, of a wholly engrossing character; but such as the duties were, he regularly and punctually fulfilled them, never receiving more than the very modest stipend of a curate, in addition to the slender income which remained after he had paid his deputy in his little college living of Moreton Pinkney.

Meanwhile the brothers at Selborne continued their improvements. On February 4th, 1759, Mulso writes:—

"In what manner to answer your last I am at a loss. We have no towering Hills, no elegant nests to copy such as I

* It held another preferment, however, shortly afterwards.

found enclosed in yours. Neither am I painter enough to give you so just an idea of them, as you I believe have conveyed to me of your Hermitage, by the handsome performance of Miss Culverton. But indeed you have shown a right picturesque imagination in the choice of the motto,* in which without the scratch under the last words I could have found, not only your poetical Fancy, but your filial Piety."

And on March 23rd, 1759—

"Every fine day makes us think of your alterations and the beauties of Selborne. Mrs. Mulso longs to see the Hermitage, the *opus operatum* of Harry."

The *Garden Kalendar* records:—

"March 31st Finished a bastion, and Ha ha fenced with sharpened piles, in the vista from Baker's Hill to the great mead: and a conical mount† about six-feet diameter at top, and five high, at the bottom of the great mead. Mount about eight days' work, Ha ha about sixteen."

In April Gilbert White paid his usual Easter visit to Oxford, the second since the year (1759) began, having gone thither for about ten days at the beginning of January, no doubt to see the Provost upon the business now settled.

* Inscribed over the door of the Hermitage were the lines from Milton's 'Il Penseroso':—

> "May at last my weary age
> Find out the peaceful Hermitage,
> The hairy gown and mossy cell,
> Where I may sit and rightly spell
> Of every star that heaven doth shew,
> And every herb that sips the dew;
> Till old experience do attain
> To something like prophetic strain."

† Later this mount received the addition of a "Winepipe" set upon it.

On his return his friend Mulso writes on May 7th, 1759:—

"I am pleased that you have got back from Oxford with a mind impregnated with Poetry, as in former days, and not troubled with Party and Contention. You brought back our old happy Feels over Milton, by those few words 'Now my task is smoothly done,' and I congratulate that placidity and academic turn of your mind."

During this spring the following letter was received by Gilbert White from his brother John, who had by this time been resident at Gibraltar about three years, as a military chaplain. It should be mentioned that John White's earlier career had been a subject of considerable anxiety to his family, both while at Oxford, and subsequently. Gilbert White had once visited Oxford on his affairs, and seems to have always acted the part of a kind brother.

Gibraltar, Jan. 10, 1759.

Dear Brother,—Your kind letter dated Oct. 26 was dated again on the Cover by my B^r^[other] B[enjamin] Nov. 17, & came to my hands, by the Cadiz Mail Dec. 13th I had been struck about a month before by the melancholy news of our dear and honoured Father's Death in the publick papers, as I was casting my eye over them in a Coffee House. I yet entertained some hopes that it might possibly be a groundless report, till I saw it afresh confirmed the next post in the Westm^r^ Journal. My grief & concern at this sad event was much increased by the long silence of all my relations in England; for the last letter I had been favoured with bears date Dec. 8, 1756, from B^r^ B, and the only ones I had ever received before were, one from my good Father, dated Oct. 16, 1756, & one from sister Reb. & B^r Harry written about the same time. You will judge how any man must be affected to have so long & painful a suspense broken by so afflicting news. The favourable circumstances which you say attended the departure of so valuable and excellent a parent ought however to afford every serious survivor no small consolation. Such a calm & unterrified resignation of his soul into the Hands of his merciful Creator must be looked upon as a blessed forerunner of the fullness of joy.

God forbid I should think of taking amiss the affectionate manner in which you express yourself relating to the perplext & uncreditable situation of my affairs at my departure from England. It is what has given me inexpressible anxiety & discontent. An imprudent and uncautious procedure at my first stepping into the world embarassed me not a little, & gradually involved me in unthought of distresses, the particulars of which I am sorry to conceal from you & ashamed to declare them. In regard to my expectations from the decease of my Father I cannot but rest in perfect satisfaction at the execution of his Will's being left in so unexceptionable hands. As soon as time will allow I must beg the favour of some farther information of the state of our affairs. For I very sincerely approve of your advice for the speedy discharge of my Debts, especially those for money borrowed. I do not clearly understand yet, how & to what extent the expenses of my education are to be estimated, but would be thankful for some intelligence on this Head. If the whole course of my maintenance & education is to be included, I fear the Ballance of my Acc^t will fall deplorably short, according to the conclusions to be made from your letter. But if it please God that matters turn out tolerably favourable, & you shall find yourself enabled to answer the under-mentioned Demands on my Acc^t, it will be an unspeakable relief to my Conscience, and I hope some recovery of my reputation. Be assured Dear B[r] that nothing but distress extremely pressing (I will say nothing at present of its cause) could have compelled me to trespass in the manner I have done on the friendliness & patience of these good people; and the loss of their esteem gives me perpetual uneasiness. An expectation of relief from my Father flattered me with a view of requiting these obligations soon, till he affectionately informed me how little it was in his power. My only income then of £40 p[r] ann. would contribute nothing thereto & I chose rather to suffer in their good-will, & good opinion for what was already past, than to amuse them with idle & useless promises & protestations. . . . I likewise acknowledge myself indebted in the sum of £10 10 0 to M[r] Bristowe, but as he thought proper, not only to furnish me with the s[d] sum in bad & diminished Coin, but also to threaten me with his Attorney in a very peremptory & insulting manner, he must give me leave to postpone settling matters with him, till providence has enabled me to do it with others. I am considerably indebted to M[r] Henry & M[r] E[d] Woods, & would be glad they would transmit to me an acc[t] of their claims. I know it is expected that I must needs save money out of my present income in Gibralter. I allow the necessity of it, tho' I may not yet have quite answered the expectations of my friends. For I find that to a person accustomed to too much remissness of accounts in his expenses, it is no easy lesson to learn frugality & economy. However, I can assure you that since my arrival here, I have by no means lived up to my Income, which I look upon as no inconsiderable point gained, but am somewhat beforehand, & hope by God's Blessing & a continuance of my present favourable situation to extricate myself from these Calamities & to evidence the honesty & integrity of my intentions. . . . I find my mechanical Talents of infinite use in this place. I have made my apartments, with my own hands, extraordinary commodious. I see by your letter that my long story about Gib[r] came to my poor Father's hands. Since I wrote that, I am removed to the most northern part of the Town—near the Grand Battery, where I have a room of a large store-house, 24 f[t] by 28. This I have divided into Dining-room, Bed-room, & study, & have plenty of cellar room, etc., in a large stair-case. My prospect from my windows is rather more elegant than the other was. I was much pleased with your description of the Hermitage, & hope some time or other to view that delightful spot once more. I beg to hear if you intend to continue the occupier of those beautiful environs of my Father's house; or if the sweet shades & solitudes of Baker's Hill, etc., are to fall a prey to the merciless Hands of the Farmer & Hedger. I flatter myself I could make some pleasing additions to your late Improvements in the fortification-way. We, who maintain ourselves here by Violence and Defiance, breathe nothing but War and Hostilities. But I fear things of this sort may look a little preposterous in the peaceful plains of Selborn & Newton; tho' the Militia Bill may perhaps somewhat reconcile them.

The end of this letter is missing.

On June 8th, 1759, Mulso wrote :—

"I have been looking over Duncombe's Translation of Horace; he has been so kind as to give us most of the Satires and Epistles in blank Lines, for I will not call them Verse. I cannot but say that you did Him great Injustice, but yourself great Justice, in not letting me insert your Imitations amongst them. They would have been amongst the things that, if he knows himself, he *desperat tractata nitescere posse*, and so leaves them to halt in Blank, but they would have ornamented the Book. . . . Mrs. Mulso is not angry at your applying to her what was designed for a Horse, for she knows that it is your favourite animal."

In the autumn of this year (1759) Harry White was presented by the Lord Chancellor to the living of Tidworth, near Marlborough, Wilts; and on December 6th, 1759, John Mulso was able to apprise his friend of his imminent removal to the valuable incumbency of Thornhill, near Wakefield. After referring to the drawback of distance from friends, etc., he proceeds—

"But you my dear Gil, who are an economist will give an immediate loose to joy for me on this occasion; and will you never see Yorkshire when you have so true a friend there? I will send you my best Hunter (for I shall be a jolly Dog;) or take any other method of conveying to me a Friend the thoughts of whom seem to aggravate the distance."

On December 21st, 1759, he inquires of his friend—

"You are an Inquisitive man, and fall into the company of many people; let me know therefore which is my road to this same Thornhill? . . . We have great faith in your Topography, as if in fact you had been everywhere. . . . I did not think that you would have increased my damp at the thought of going so far, by supposing that we should be more separated than ever. What are a few miles more to you on a cheap Road? Lyndon must be half way or more, and a horse can go anywhere. It is impossible, but that your very curiosity must bring you to see us. What! the brave Mercians, and the Castella Brigantum left unseen by a man who will send miles for a huge stone or a knarly root of a tree? Nothing but death or marriage will make me believe it. . . ."

Though neither of these events was the obstacle, nevertheless his friend never saw Yorkshire. On October 17th, 1759, he received the last payment "in full" from Mr. Etty for serving Selborne cure; and in November he set out for his brother-in-law's house at Lyndon, *viâ* London. Mr. Barker shared with his brother-in-law a distinct taste for the observation and record of facts in natural history. A MS. diary of his—now, through the kindness of a cousin, the property of the present writer—contains, besides observations in astronomy, notes on the measurements of trees (including the great yew tree in Selborne Churchyard), a record of the first appearance and disappearance of migratory birds, and the breaking into leaf and flowering of trees at Lyndon for the series of years 1736–1801. This diary is mentioned in the original letter to Pennant of August 17th, 1768 (Letter XIX.), from Gilbert White:—

"Now I present you with a paper of remarks from Thomas Barker Esqr of Lyndon Hall in Rutland, a gent: who married one of my Sisters. In it you will find, I think, a curious register, kept by himself for 32 years, relative to the coming and departure of birds of passage. If you find anything in it, or among ye rest of the observations worthy y^r^ notice, you are wellcome, he says, to make what use you please of any of them."

After leaving Rutland, Gilbert White returned to London to stay there with his brother Thomas. Returning to Selborne, he records the fact on

May 17th, 1760, in the *Garden Kalendar*, "after six months' absence in Lyndon and London," the only long holiday of his clerical life, and the last considerable absence from Selborne. Meanwhile his friend at Thornhill wrote on March 2nd, 1760:—

"You are impatient to have me say something of Thornhill: but I shall say nothing at present, that I may leave the greater impetus to come to us. Only that here are such great capacities for improvement, that it is absolutely necessary that you should come hither. . . . The roads are worse than yours for a carriage. We have some scenes almost Selbournian, but not your turf or soft Woods or wild Hangers."

CHAPTER VII.

PROBABLY Gilbert White was by this time beginning to realise that Selborne would always be his residence, since the *Garden Kalendar* records in June, 1760, that he "agreed with John Wells to purchase the upper part of Lassam's orchard," which was added to the premises of "The Wakes." A little later he writes:—

"Tull and John * are busy every day in grubbing, paring, and burning the new purchased garden."

On August 11th, 1760, Mulso writes:—

"If you have the same course of weather we have here, you have not yet dined in The Hermitage. The image of that place, and the pleasures we enjoyed from the circle that belonged to it, are so strong in our minds that we feel your disappointment as if the case was our own. I look up to the picture of it (it hangs over my study door) and fill myself with many a friendly thought. I enjoy your new purchase extremely; I see the alteration, I could draw it now, but I

* Robin Tull was occasionally employed in the garden. An entry of "John's livery" in 1760 shows that John (? Lassam) was now the regular manservant.

want paint for the deep verdure of the βαθυκόλπος meadow; and I remember the old Trees, that would embarrass the expressions of our Garrick."

The letter concludes with congratulations to Mr. Etty, "and to yourself on a new neighbour"—presumably Mrs. Etty.

Later, on November 3rd, 1760, he writes:—

"I enjoyed your 'Day at The Hermitage' very much."

On January 13th, 1761, Mulso apprises his friend that—

"The long expected union of my brother and sister Mulso,* and of my brother and sister Chapone,† has at length taken place. To these brides and bridegrooms I know you will give your good wishes, that, as they have long wished for *this happy state* (I don't know whether I speak to be understood by you who continue an old Batchelor) they may long continue happy in it."

He goes on to say—

"I am glad your Oxford journey ended so much better than it began, and that the Provost and you begin to have your old Feels for one another, such as you had before Competitions divided you. I hope to find in the long run, that with the help of the public coalition of Parties, and his own interest Dr Musgrave will prove a very serviceable Head of that House, and, as I know you have the good of the Foundation at heart, it will make you forget what was once disagreeable in his election."

It will have been amply apparent that Miss Mulso's marriage was not an event likely to occasion

* Thomas Mulso and Miss Prescott. † *Née* Miss Hester Mulso.

other feelings in Gilbert White than those of friendly interest; but whatever his feelings may have been, his interest in his garden did not abate.

The following entry in the *Garden Kalendar* refers to an improvement which remains to the present time:—

"Jan. 24 [1761]. Long the mason finished the dry wall of the Ha-ha in the new garden which is built of blue rags, so massy that it is supposed to contain double the quantity of stone usual in such walls. Several stones reach into the bank 20 inches. The wall was intended to be 4 ft. and an half high. But the labourers in sinking the ditch on inclining ground mistook the level, especially about the angle, so that at that part to bring it to a level it is 5 ft. 8 inch high, and 4 ft. 6 inch at the ends, an excellent fence against the mead, and so well fastened into the clay bank that it looks likely to last a long while."

"Feb. 5. Levelled the Terrass and new walks so far that they will want but very small amendments before they are turfed."

An entry taken from his account-book at this time refutes the supposition that Gilbert White was living as a rich man at Selborne.

	£	s.	d.
"Feb. 18 [1761] Paid sisters's housekeeping from July 2 to Feb. 19 . . .	20	02	04
[At this time he paid for a year's rent of his house and orchard]	05	07	00."

A little later a further piece of ground was bought. Mulso, referring to this on April 20th, 1761, congratulates his friend on "having got the *angulus iste.*

I observe you are not for us this spring. There is a wall between us, but no gulf, indeed we long for the sight of you."

In April he paid "Fine Copy and Fees for the new garden" to Magdalen College, Oxford, the owner of the manor of Selborne.

In May, 1761, the *Garden Kalendar* contains a long entry, commencing—

"May 20—My brother Tho and I went down with a spade to examine into the nature of those animals that make that chearful shrill cry all the summer months in many parts of the south of England."

This entry, subsequently re-written, became the forty-sixth letter to Barrington. The *Garden Kalendar* continues—

"June 21—Discovered a curious orchis in the hollow shady part of Newton-lane, just beyond the Cross. It is the *orchis alba bifolia minor calcari oblongo;* grew with a very long stem, and has been in flower some weeks. I brought away the flower, and marked the root, intending to transplant it into the garden, when the leaves are withered."

Later there is mention of another improvement which remains, in part at least, to this day in the garden of The Wakes.

"July 25. Finished my fruit wall, coping the two returns at the end with stones of a sandy nature out of the old Priory. The coping bricks were full of flaws and cracks, being made of earth not well prepared, and instead of overhanging the wall, came but just flush with it: however, by using six that were broken-ended, we had just enough, and they may lie on the wall many years."

In November was purchased "a large Portland dial-post and slab from Andover," and "a brass dial plate" for a sundial.

On September 7th, 1761, Mulso writes from his Yorkshire living, adjuring his friend to write more frequently—

"By *Hercules** and by the *Dragon* on the Cynic Tub, I swear, I ought not to be forgotten upon Selbournian Ground. . . . I desire that you would make a better use of this winter than you did of the last. How can *Mice* [Miss White] and you sit one on one side and t'other on the other side of the fire, when you ought to be both at the great table in the middle of the room, she writing to her lover, and you to your friend? I mean after your Journal book is filled for the day, for I do not expect to be served first."

In the autumn of this year Mrs. Chapone lost her husband, after a short married life of ten months. On October 29th, 1761, Mulso communicated his sister's "irreparable loss" to his friend, and continued—

"Your employment has of late been of a more cheerful nature; the entertainment of your friends, in which you shine; and in the disposal of your sister. . . . Her temper and conduct must, I am sure, very much endear her to the man whom she has honoured with her Choice. But what will you do for a housekeeper? or have you looked till you have found one who will more than supply her place? . . . I long to see your alterations and improvements at Selborne, tho' I have a pretty strong idea of them. . . . I thank you

* In 1758 a "Board-statue of Hercules" appears in his account-book. Cf. "The Invitation"—

"Or with the mimic statue trap the sight."

for recording our friendship with so much warmth. It is, in my esteem, one of those blessings of God, which partake of his nature, and are therefore inestimable."

On November 17th, 1761, Miss Rebecca White married her cousin, Mr. Henry Woods, whose father resided at Chilgrove, near Chichester, and went to live in London; leaving her brother, the last of the once numerous family, at Selborne. Apparently he solaced himself with continued improvements in his garden. On December 30th he records in the *Garden Kalendar*—

"By the negligence of Murdoch Middleton my wall-trees never came till the 26th. They are in general good trees, and were planted, (considering the wetness of the season,) in good condition; and in the following order, beginning from the terrass.

"Breda apricot: Sweetwater vine: Roman nectar: Mr. Snooke's black-cluster vine: Roman Nect. white Muscadine vine: Mrs. Sn. Newington nectar: Mur: Middleton's Sweet-water vine: Nobless-peach: Mr. Sn. white Muscadine vine: Nobless-peach: John Hale's two Passion flowers, one at each end of the wall.

"Planted two Cistus's in Mr. Etty's dry garden; and a Phlomis, and an Halimus in my own."

In December, 1761, a journey was made to the vicarage of Moreton Pinkney. Referring to this, Mulso writes on February 5th, 1762:—

"We wish you joy of your sister's wedding, tho' it is now almost an old history; as I remember I did before it was solemnised. But what a Hussar Parson do you still continue to be? and how did you giddy me and hurry me along with your account of your Journey, as bad as Tristram Shandy's

Calculation of the Quantum of Genius thro' all the northern nations. It is well for me that you are set down quiet at Selborne, where I suppose the coming of the Spring will confine you till you have set everything in order for its summer Perfection. As soon as that is done, I think I hardly know you, if you do not set out on a Ramble."

Since Michaelmas, 1761, Gilbert White had been curate to Mr. Roman, who had succeeded Dr. Hales as rector (non-resident) of the adjoining parish of Faringdon; a curacy which he held for five-and-twenty years, until the death of Mr. Etty, and the non-residence of his successor, gave him the opportunity of becoming for the third time curate of his native village.

In March, 1762, the *Garden Kalendar* contains a careful entry of the order in which some new pear trees were planted in the "new garden." The names may now be of some little interest. They are: "Chaumontelle," "Virgoleuse," "Crasane," "Doyenne." "St. Germain," "Brown Bury," "Autumn Burgamot," and "Swan's Egg." A "Le Royal" and a "Queen Claudia" plum (greengage) were also planted. April, 1762, saw Gilbert White set out for Tidworth, which Harry White was just about to leave, since he had been presented to the (Chancellor's) living of Fyfield, near Andover.

Mulso writes on April 18th, 1762:—

"I hope my brother Rector of Fifield will change his for a better under the Patronage of my Lord Chancellor; in the meantime I sincerely wish him joy of the present

Preferment. How agreeable do you men of Taste make every Place you come at! Your description of Tidworth made me think of your prints of the antediluvian ages; and they conveyed simple and rural images, which amuse and deceive the fancy; but yet Tidworth may be exceeded, and perhaps none since Adam have seen so fine a scene as Mr. Morris's Gardens. . . . I like your verses extremely, the thought is happy, and the execution entirely in the vein of Ovid. The fourth line I have some objections to. . . . I am pleased to catch you versifying. You will never be old"—

a true enough remark; few men retain to their old age such a fresh and healthy enjoyment of life as did the person addressed.

The summer of 1762 passed away, and though he had received frequent and pressing invitations to visit his old friends in Yorkshire, the Naturalist was either unable or unwilling to leave home and his curacy for any long period. Moreover, his uncle at Bradley, the Rev. Charles White, was at this time in greatly failing health, and this kept the nephew near him. Meanwhile the improvements and alterations in the garden at Selborne continued. His friend writes on October 9th, 1762:—

"You tell me of an alcove at the end of your Terrace. Which is your Terrace? for you had no walk of that denomination when I saw Selborne. Is it the north side of Baker's Hill?* or is it near the other Bench, where the opening and new Bastion was, facing the Cynic Tub? Clear me up: for I am lost in the grandeur of your outlets, and the multiplicity of your improvements."

* Baker's Hill is the small elevation in the little park belonging to The Wakes, on the left hand, looking towards the Hanger from the house.

In November, 1762, the *Garden Kalendar* notes—

"Shut up the alcove with straw doors for ye winter; and took in ye urns."

And in December—

"Finished a paved footpath from the Butcher's shop to the Blacksmith's. . . . It cost just one pound."

Early in March, 1763, Mr. Charles White died. He was buried in the chancel of Selborne Church. It appears that his nephew applied to Lord Chancellor Henley—in whose private patronage the rectory of Bradley lay—for this living, but met with a refusal. Had he received this small living he would not, judging from a remark of John Mulso's, have discontinued residence at Selborne; especially as he now succeeded his uncle as owner of the house which he occupied, and also received from him some small properties in and near Selborne.

A certain importance now attaches to the residence of Gilbert White at Selborne, and it may be mentioned that there is no bequest of The Wakes in Charles White's will, though he undoubtedly owned and received rent for it, according to his nephew's account-book, in 1762. This house was thus bequeathed by Gilbert White, Vicar of Selborne, to his daughter Elizabeth (Mrs. Charles White), "and her heirs for ever after the terme which I have left it to my wife for I give likewise my copyhold estate in Selborne which followeth viz—my house and orchard in Selborne Street late Wakes."

Why Charles White omitted to bequeath The Wakes to his nephew Gilbert is not clear, since he left him "the orchard or garden, now in his [Gilbert White's] father's possession,* and adjoining to his father's house." He mentions, however, respecting some other property formerly Mrs. Charles White's, "supposing that the said Gilbert White will enjoy [it] after my death as heir at Law to his aunt my late dear wife."

Mr. Bell, who owned the property, in his edition of "The Natural History and Antiquities of Selborne," vol. i. p. xxvii., states positively that Gilbert White "became the actual possessor of the property only on the decease of his uncle Charles, the Rector of Bradley, in 1763. The deed† is thus endorsed in Gilbert's handwriting—

'Copy for the Wakes my Dwellinghouse,' etc."

Referring to the refusal of the living, Mulso comments thus in an undated letter, apparently written about April, 1763, upon the situation:—

"I am sorry that the Patron refused you what you asked for, as it would have brought your matters within Compass, and you might still have resided at your beloved Selbourne.... You are now arrived at that happiest of human states, Independence. For though you may be glad to have an addition to your Fortune, especially in lieu of your Fellowship if you find yourself inclined to marry, yet as you have all in your

* The will was made in 1753.

† Presumably his copy of the admission to the copyhold tenure of the property on the Manor Rolls.

own power, without a necessity of attending and soliciting anybody, you are in an enviable situation by the greatest part of mankind. By the last paragraph in your paper we have hopes of availing ourselves of your freedom some time this year, when your affairs are put into train."

Later, on July 28th, 1763, Mulso continues upon the matter of the living. Referring to his uncle, the Bishop, who had just paid him a visit at his Yorkshire living, Thornhill, he remarks—

"Among other subjects of discourse, you was one. I showed my Lord your last letter, from which I imagine that you did not purpose any other good and present effect should arise, than a general disposition to serve you at some future happy opportunity. The Bishop, who is on all sides beset, did not chuse to enter into any sort of promise of this kind; but, as I had before returned him my particular thanks, as well as yours, for his interposition with my L[d] C——,* and as I stood for you, as far as I thought would be serviceable to you, I cannot think you will have the worse chance with him for this visit.

"I found him a little inclined to blame you for not having managed your cards with L[d] C—— better. But tho' he blamed you to me, he defended you to him. As I could not but express a great surprise at your not being better in that Person's good graces, (not having ever suspected it), I asked my Lord if he knew the cause: he said very well; it was given him by himself the first time he spoke to Him about you. You did *not* vote for the Bishop of Durham at Oxford.†

* Lord Chancellor Henley, afterwards Earl of Northington, was the friend and patron of Charles White, whom his family had presented to the livings of Swarraton and Bradley, Hants, which were in their private patronage.

† There was a contest for the Chancellorship of the University of Oxford on May 4th, 1750, between the Earl of Westmoreland and Richard Trevor, Bishop of Durham, who had been a Canon of Christ Church, Oxford. The former was elected by 321 to 200 votes.

My uncle being better informed in this affair before the next day at the House, went up to him and told him that you *did* vote for him; to which he answered to this purpose, 'that it was in so lingering cold and disobliging a manner, that he could not but believe you disinclined to any services of that nature.' Now what my uncle thought faulty, was, that knowing his Pride and expectations, you did not enter with a more sanguine shew into his intentions, if you entered into them at all. I answered for you, that you had taken the journey on Purpose, that you thought *voting* was all that my Lord wanted from you, which you had done: and that if you did not come in till late, he who was an Oxford man, could not but know how exceedingly disagreeable it was to stand the Brunt of a majority in your College of the opposite Party."

In short, his late uncle's living was lost to the applicant because he was not quite a good enough courtier.

In the summer of 1763 the *Garden Kalendar* notes that—

"July 13 Mr. Thos Mulso & Lady & Mr. Edwd Mulso and Miss Harriot Baker came to visit me."

These were not the only visitors at Selborne at this time, since three young ladies, the Misses Anna, Catharine, and Philadelphia Battie, the handsome daughters and co-heiresses of Dr. Battie, a very eminent London physician,* who were cousins of Mrs. Etty (*née* Littleton), were at this time spending some weeks at the vicarage.

* Dr. Battie was President of the Royal College of Physicians of London in 1764.

Walker & Cockerell, ph. sc.

Catharine Battie

Writing on July 12th John Mulso says :—

"How busy will you be when this reaches you, in showing out your delightful Places? My sister Mulso, if she has strength to reach them, has a soul to enjoy them; my brother's picturesque genius will find food. Yes, I see you upon the area of the Hermitage, the arm extended and the finger pointing out the happiest lights and Shades of the prospect. I see you under the beeches of the Lythe,* You are in more soft and mild attitudes, a sort of Pastoral spirit possesses you; you hardly want to look over the blue forest, so contented are you in your green recess. . . . What a cold Batchelour are you! So many Beauties, and so many Thousands, and *Integer laudas?* Very fine! abominable I mean. As to poor Harry Tinderbox, I pity his liver. Desire my brother to use his pen, as well as his pencil, and let us be the better for it. . . ."

Miss Catharine Battie, at this time in her twentieth year, kept a little journal of her summer visit at Selborne, which the writer of this memoir, her great-grandson, ventures to think will be found of sufficient interest to excuse its being given in full.

A little Journal of some of the Happiest days I have had in The happy Valley in the year 1763.

The 4 of June 1763 dined in the Hermitage the company Mr. White Mr.† & Mrs. Yalden Uncle Richard Miss Jenny White‡ Mr. & Mrs. Etty my Sisters & self, Mrs. Snooke§ came up to Tea with us had a very agreable day.

* Common land, near the stream, at Selborne.
† Vicar of Newton Valence.
‡ Daughter of Benjamin White.
§ Of Ringmer, Gilbert White's aunt, now lately a widow.

The 10 of June at Hackwood* the Seat of the Duke of Bolton's it is a sweet pretty place walk'd about the Woods all the morning then went to see Basingstoke Church after dinner went back to Hackwood to see the Kitchen Garden then we remounted our horses but Kitty being tired of her Rider soon laid me down softly upon a fine green Carpet. Mr. White rode the rest of the way home, we had a very pleasant day

22 of June in the afternoon Mr. White Mr. Harry White Mrs. Snooke & Mrs. Woods drank tea here; Mr. H. White & Nancy sung & play'd at nine o clock Dr. Stebbing† & Mr. Airson‡ came. Went to bed between twelve & one o Clock was very merry after supper the next day being Mr. & Mrs. Ettys' wedding day we kept it with mirth & jollity. The morn was spent at the Harpsicord a Ball at night began minuets at half an hour after seven then danced country dances till near eleven went to supper after supper sat some time sung laugh't talk'd & then went to dancing again danced till 3 in the morn: at half an hour after four the company all went away we danced 30, danced, never had I such a dance in my life before nor ever shall I have such a one again I believe.

The Company at the Ball were

Mr. Etty & Mrs. Yalden
Dr. Stebbing† & Mrs. Harrison
Mr. Airson‡ & Miss Battie
Mr. Yalden & Miss Higgens
Mr. Harry White & Miss Kitty Battie
Mr. Foxcroft & Miss Newlin§
Mr. Harrison & Miss Philly Battie

* Near Basingstoke.

† Rector of Beaconsfield 1768–1801, at this time engaged to Miss Littleton, Mrs. Etty's sister.

‡ A clergyman residing near Winchester.

§ Probably daughter of Richard Newlin, Rector of Greatham.

The Spectators were

Mr. White Mrs. Snooke
Mrs. Woods Miss Littleton
& Mrs. Etty.

June 24 got up at ten in very good spirits (who can be otherwise in this D^{r} [dear] place) after Breakfast Mr. Whites came in at eleven o Clock went to Church at two we ascended the zigzag, up to the enchanting spot,* where we dined after dinner we went to the Tent† here we sat till 6 when we went back to the Hermitage to tea in the middle of tea we had a visit from the old Hermit‡ his appearance made me start he sat some time with us & then went away after tea we went into the Woods return'd to the Hermitage to see it by Lamp light it look'd sweetly indeed. Never shall I forget the happiness of this day which exceeded any I ever had in all my Life, sweet Hermitage agreeable Company fine day good spirits all combin'd to make it of all days the most agreeable.

Sat 25 after Breakfast Dr. Stebbing went away. Mr. Harry White came in to take his leave of us being in a great hurry to go but the 3 sorceresses Nanny Kitty & Delphy so bewitched him that he did not go till four in the afternoon.

Monday July 11 Dr. Stebbing came Mr. White drank tea with us after tea Mr. White Mrs. Woods & myself went to the Hermitage.

Tuesday 12 bad day nothing done worth taking any notice of.

Wed: 13 in the morning Dr. Stebbing, Miss Littleton & Delphy went out a Horseback Mr. Budd came to dinner

* The Hermitage.

† An ornamental tent is shown in the large view of Selborne in the first and third (quarto) editions of "The Natural History of Selborne."

‡ Harry White personated the Hermit.

after dinner Mr. Harry White came in we sung & Play'd till tea time, after tea he came in again & Play'd till supper time. Mr. & Mrs. Mulso Mr. E. Mulso and Miss Baker came to Mr. White's.

Thursday 14 drank tea & spent the evening at Mr. White's, after tea singing playing & dancing, after supper the two Mr. Mulso's & Mr. Harry White sung catches never laught so much in all my life.

Friday 15 Mr. Harry White breakfasted with us all the morn was spent at the Harpsicord in the afternoon all the family from over the way came here except Mrs. Mulso the afternoon was spent in a most delightful manner with Correlli & Handel at ten went to supper at one in the morn the gentlemen & us changed Caps & wigs several minuets were danced Dr. Stebbing danced a charming one.

Sat. 16 Dr. Stebbing went about one o Clock Mr. Cane* & Mr. Henry White came on Horseback to the Door to take leave of us determined not to come in but he soon broke thro' his resolution & dismounted came in sang 3 songs & then took his leave.†

Sunday 17 went to Church twice drank tea at Mr. White's the company agreeable as usual.

Monday 18 Mr. E. Mulso here all the morn. Sister & Mr. M. sung & Play'd drank tea at home by ourselves afterwards went to Mr. White's.

* The Rev. Basil Cane, Gilbert White's first cousin.

† The following note, which Miss C. Battie carefully preserved, apparently refers to some bet between her and Gilbert White upon this delayed departure :—

MADAM,—I make a point of paying my debts of Honour as soon as possible: but at the same time can't help remarking that it was not a fair wager. For it is plain, by some art magic best known to yourself, you have not only a power of detaining men that ought to be going; but also of keeping those away that ought to come. In the whole it is best that I have been the loser, as it would not be safe in all appearance to receive even so much as a pin from your Hands.—I am, with many a x x x x x x x and many a Pater noster and Ave Maria, GIL. WHITE.

Tuesday 19 after breakfast Mr. W. came in to ask us to go out a Riding we drest and went over to his house but the weather grew so bad that it prevented our going We spent the morn together with much mirth & cheerfulness we were all weigh'd to see how much we were worth. I weigh 134 lb oh monstrous afterwards we were all measured came home to dinner in the afternoon Mr. White the Mulsos etc. etc. came here with work singing & Playing we spent a very agreeable evening mem. Charles* was put into long [*sic*] Coats look'd vastly pretty.

Wed 20 after breakfast Mrs. M. Mrs. Woods & Miss B. came in Miss B. Philly and self & the two Mr. Mulsos went up to the Dr Hermitage Mr. T Mulso gave us a discourse upon Natural Phylosophy & Astronomy we work'd he read some of Thomsons Seasons we walk'd in the Woods & then came home to dinner I hope I edify'd by his sensible discourse Mr. E. M. & Mr. White drank tea here afterwards we went out a Riding I rode double for the first time rode upon the sweet Commons & in the High Wood call'd at Newton, it was a most delightful evening I hope we are going to have fine weather.

Thursday 21. after Breakfast went into the Hay field toss'd the hay about a little then went to Mr. White's sat in the Alcove spent the morn most delightfully Mr. T. Mulso read Thomson & at Two came home to dinner at 6 we met again to walk went up to the sweet Hermitage sat viewing its various beauteous [? views] some time then walk round the wood back to the Hermitage, Mr. White read us an acrostick made upon Nanny. Miss Baker & I found a stone upon the Common which we carried to the Hermitage & placed it there as a memorial of our fondness for that place.

* Son of Mr. and Mrs. Etty. When he grew up he entered the naval service of the East India Company, and was the "young gentleman" who brought home the Chinese dogs mentioned in Letter LVIII., to Barrington, of "The Natural History of Selborne."

Friday 22. Work'd in the Garden till one o Clock H. Baker with us from one to two musick. Mr. E. Mulso dined with us sat in the garden after dinner at five went to Mr. White's drank tea under the Tent in his sweet fields all the Newton family there, walk up in the evening to the Hermitage another acrostick upon Nannie.

Sat. 23 Mrs. Mulso Mr. E. M. & Miss Baker here in the morn we work'd Nanny & Mr. E. M. Play'd in the afternoon they drank tea here it Thundered & rain'd I had the head-ache very bad.

Sunday 24. went to Church twice drank tea at Mr. Whites it rain'd all the morn the evening proved fine we walk'd in his sweet fields.

Monday 25 rode in the morn. with Mr. White to Church at Faringdon came home at one Mr. Henry White & Mr. Cane came they all dined here upon a Haunch of Venison after dinner musick & singing the Newton people came down after tea we danced before supper. Mr. E. Mulso sung us a song made upon the 3 sorceresses set to the Pastorrella, after supper many songs & catches were sung at one they went away. We had a very agreeable day how is it possible to do otherwise with all these cheerful goodnatured People. We were all Shepherdesses.

Tuesday 26 went out a Horseback behind Mr. H. W. as soon as we got upon the common it began raining We stop at Newton came in like drowned rats were very merry came home in the rain behind Mr. White in the afternoon drank tea with him sisters & self had our shaddows taken.

Wed 27: after breakfast wrote musick then went over to Mr. White's to be electrified in the evening walk'd to Noar Hill oh sweet evening sure there never was anything equal to the romantickness of that D[r] D[r.] Hill, never never shall I forget Empshot & the gloomy Woods the distant Hills the South Downs the woody Hills on the right Hand, the forest the valleys oh all are heavenly almost too much for one to bear, the sight of this beauteous prospect gives me a pleasing melancholy. 6 of us walk 7 rode got back at nine.

Thursday 28 in the morn; Rode to the Barnet from whence we had a Prospect of Wilts; Hants Berks Sussex Surry, got home to dinner at two after dinner went up to the Dr Hermitage we drank tea afterwards the old Hermit came to us he told Nannie and H. Baker their fortunes after the old man had sat some time with us he retired & we adjourned to the tent where we Shepherdesses * danced; at nine the lamp was lighted, enchanting scene oh never did I see anything like it 'tis 'tis Arcadia Happy Happy Vale when shall I see thee again.

Friday 29 after Breakfast the gentlemen came over here musick all the morn dined & supt at Mr. Whites after dinner Singing & Playing & Shadows taken after Supper Catches & Songs. At one in the morn we changed Caps & Wigs with the Gentlemen & danced minuets spent a very agreeable day.

Sat 30 Mr. Henry White breakfasted with us afterwards went over the way heard the Eolian Harp there we had some of old Handel at two Mr. H. & Mr. Cane went in the afternoon we all went to Newton, very merry, Corridon dined with us upon a leg of mutton.

Sunday 31 went to Church twice Mr. White Pray'd & Preached after Church we all walk'd together in our garden *the* [*sic*] over the way drank tea here, walk'd to Hartley saw

* This costume was apparently suggested by Gilbert White, since Miss Battie preserved the following verses in his handwriting:—

"Gilbert, a meddling, luckless swain
Must alter lady's dresses,
To dapper Hats, and tuck'd up train,
And flower-enwoven tresses.

"But now the Lout with loss of heart
Must for his rashness pay;
He rues for tamp'ring with a dart
Too prompt before to slay!"

the House & Park the House stands deplorably some views from the Park are very fine.

Monday August 1st after Breakfast we went out a Riding at Newton Gate we were all seized with a sudden panick & dismounted sent our Horses home & walk'd into the High Wood from there to the Hermitage where we sat for some time; as we were coming down the Zig-Zag what should we spy in Punt field but Strephon & Collin return'd oh the Poor bewitched men. Went home to dinner afterwards we went into the Hayfield underneath our garden then went over the way to tea after tea walk to the Priory had a most delightful walk thro' Woods & hay fields.

Tuesday 2nd after Breakfast musick till dinner time Corrydon & Strephon here Mr. White etc. drank tea here & supt danced & sung merry & happy as usual.

Wed 3 before breakfast Strephon came in and presented me with an elegy upon my departure * after breakfast went over to Mr. White's for a little while then came back all

* Carefully preserved by the Diarist, with some "Relicts of the dear Hermitage," are the following verses addressed to her by Harry White:—

"DAPHNE'S DEPARTURE;

"AN ELEGY WRITTEN AT SELBOURN, 1763.

1.

"Too plain my Heart forbodes that fatal Hour,
When dearest Daphne leaves the happy Vale;
Let then this Breast own Love's resistless pow'r,
Nor longer strive its anguish to conceal.

2.

"In vain stern Reason bids me check my Woe,
Or stay the Current of these fruitless Tears;
For thee dear Nymph! they must for ever flow,
Fed by the fountain of thy Strephon's fears.

3.

"Ah! wretched Selbourn! what avail thy Shades
Thy lofty Hills with waving Beeches crown'd;
Their boasted Glory now for ever fades,
And endless Winter shall thy Vales surround.

Gilbert, a meddling, luckless swain,
Must alter Ladies dresses
To dapper hats, & tuck'd-up train,
And flow'r-enwoven tresses.

But now the Gout with loss of heart
Must for his rashness pay:
He rues for tamp'ring with a dart
Too prompt before to slay!

[To face p. 136, Vol. I.

came with us walk'd round the sweet Garden came into dinner no stomach to it at half an hour after one left the happy valley with a heavy heart the whole company cry'd Strephon, Corydon & Collin rid with the post-Chaise parted with poor Strephon & Collin at the Wheatsheaf with Corydon at Alton, here the scene closes the play is done the pleasing dream is oer & tomorrow I must awake & find myself in London. Adieu happy Vale enchanting Hermitage much loved stump beauteous Hanger sweet Lythe

4.

"How sweetly I, at close of Summer's Day,
While thy dear presence blessed these happy Bowers,
Could lost in rapture with my Daphne stray,
Or in soft converse pass the fleeting Hours.

5.

"The mazy Zigzag—or th' inchanting Cell—
But Scenes like these no longer chear my Eyes;
Nor shining Lake, nor wild romantic Dell,
Nor craggy Mountains soaring to the Skies.

6.

"The hoary Hermit in his calm Retreat,
No longer safe from her resistless Charms;
With trembling Hand, dim Eye, and faultering Feet,
Sighs out his Dotage o'er her snowy Arms!

7.

"Alas! how changed! already seems the Scene,
Where Pleasure erst triumphant held her Sway;
No more her jocund Train with Smile serene
Shall chear each mournful—each long lingering Day.

8.

"Then, let me, Fairest! fly this once lov'd Place,
Forlorn to vent my sighs in hopeless Grief;
E'en thy dear Image from my Soul efface,
And in Oblivion seek a sad relief.

9.

"But see!—my Charmer drops the pitying Tear!
Oh may she prove indulgent to my Pain!
This Heart that loves her with a Flame sincere,
Shall then its wonted Chearfulness regain.

"STREPHON."

to all I bid adieu with grateful thanks may the Woods flourish may no mischievous Boy hurt the little Nest may all the good inhabitants have as many happy days there as I have had 'tis the sincere wish of

DAPHNE.

Though London seemed to this young lady such a dismal residence after Selborne, it may be recorded that the bow-window of the back drawing-room of her father's house in Great Russell Street, at the corner of Montague Street, the site of which has long been absorbed by the British Museum buildings, at this time commanded an uninterrupted view of the Highgate and Hampstead woods. A few years afterwards the eldest of the three sisters married Admiral Sir George Young; the second, Catharine, became in 1771 the wife of John Rashleigh, of Penquite, Fowey; son of Jonathan Rashleigh, of Menabilly, Cornwall; and the youngest, Philadelphia, married Mr., afterwards Sir, John Call, Bart.

On July 28th, 1763, John Mulso writes:—

"I received a letter last night from my dear brother Ned. If he is still with you I beg you to thank him for it. I gather from it that you find your guests in a variety of entertainments, and keep them constantly employed in gallantry. They are now reduced, by the departure of the *Sorceresses*, to the elegiac strain; and must at least hang verses on the Beeches of the Hanger and the Noar, if they do not serve themselves in the same way."

Though he did not hang either himself or his verses on the Selborne Woods, Gilbert White soon afterwards, on November 1st, 1763, composed his now well-known verses "On Selborne Hanger, a Winter Piece" which he dedicated "To the Miss Batties," wherein he celebrated

> . . . "The scene that late with rapture rang,
> Where Delphy danced and gentle Anna sang,
> And on her stump reclined the musing Kitty sate."

The occasion of the departure of these ladies also called forth the following lines by Gilbert White, now printed for the first time:—

"KITTY'S FAREWELL TO THE STUMP BENEATH THE HERMITAGE.

> "Kitty, a grateful Girl, in doleful dump,
> On the steep cliff deplor'd her favourite stump.
> 'Shall I no more (the melting Maiden cries,)
> With thy *sweet* scenes delight my feasting eyes:
> Say, shall no more thy nearer views engage,
> The crouded Tent and swelling Hermitage;
> Shall the cowl'd Sage no more my sight beguile,
> So stout, yet so decrepit all the while;
> Wond'rous old man, in whom at once combine
> The hoary Hermit, and the young Divine?
> In dapper hat adorn'd, and pastoral dress,
> Must I forget each Sister Shepherdess;
> While amorous swains, bending with anxious care,
> Down the loose slope conduct their sliding Fair?
> Delightful stump! on thy rough lap reclin'd
> How shall I paint the transports of my mind?
> Words are but vain: come then the starting tear,
> And soften'd feature, speak my love sincere.'"

The *Garden Kalendar* records—

"July 28. Drank tea twenty of us at the Hermitage; the Miss Batties and the Mulso family contributed much to our pleasure by their singing and being dressed as shepherds and shepherdesses. It was a most elegant evening; and all parties appeared highly satisfyed. The Hermit appeared to great advantage.

"Aug. 3. Cut a fine-looking cantaleupe and sent it by the Ladies (who left Selborne this day) to Dr Battie."

On October 3rd, 1763, Mulso writes again, referring to the question of the living:—

"I read to the Bishop your letter, wherein you relate your proceeding with your Patron. . . . You need not doubt that I urged my own opinion of you to him, as far as was decent: and though not one word was said in return, I hope it will operate in Time. I heartily wish it may; that my friendship may not have been quite unfruitful to you in point of interest. . . . Your affection, which has always shown itself in every possible instance, demands a constant desire in me to please and serve you. It has been one of the blessings for which I am ever thankful to Divine Providence, and it has greatly helped to enliven and sweeten the painful passage of human life.

"That I might give you some account of my last interview with the Bishop of Winchester was the principal reason of deferring answering your last, for otherwise I should have been quicker to quiet your apprehensions of our ever suspecting that you had not done all possible to make us a visit in the north. . . .

"I am glad so many of my relations have seen Selbourne: they all express the highest pleasure in their visit. As to the consequences to the young gentlemen, it will be no great matter, especially as it purges off in Poetry: when Passion is fanciful it is not dangerous. Ned requires these Brushings, being apt to have torpid and viscous blood, if a love fit now and then did not quicken his pulses. It would not do you so much good, unless it was once to the Purpose;

for we, my friend, begin to grow into a more serious age, and to mean a little more what we profess. I beg you to get as much this winter as possible into the gay world; for it will be of prejudice to your health and spirits to employ a winter in putting on wood in a country village. At all events, and wherever, write to me often, for when even the mind alone expatiates, the body is something the better for it."

On December 7th, 1763, Mulso writes again to his friend who was visiting in Rathbone Place:—

"A little of the bustle, and talk, and variety of London is absolutely necessary for you. And if you should have any *farther knowledge* of the Miss B's, it might have rather a salutary than a dangerous effect, for it is my notion that they may be very safely taken either full or fasting. I hope the town does not disagree with you after so long an absence from it. I know you used to be sorely beset about bread and beer in London. We sit at a distance and are but little troubled here about things that wear a grave face with you; but perhaps a little of the wiser business of London may not be disagreeable to you, after having been so far from it as Arcadia. How very different! and how many people did you make happy last summer! Some indeed bear some scars; for even the sweet rosebush has its thorns. I endeavoured to laugh at some of them as mere scratches, but alas poor King Lear* was mad in his last letter, the paroxysm was upon him and he complained that he was *cut to the brain*. So I have not dared to say a word yet, till the heated imagination is grown cool. I hope the poor curates at Lurgeshall† are come to themselves again."

* Qu. Ned Mulso, "Corydon"?

† Basil Cane, "Colin," was curate of Ludgershall, Wilts, two miles distant from Harry White's, "Strephon's," living of Tidworth.

CHAPTER VIII.

That, as regards Gilbert White at least, no very deep impression had been made by the charming shepherdesses, may be gathered from a remark which John Mulso makes soon afterwards (January 6th, 1764)—

"So that I find the Miss B——s are still ladies 'that you *know but little of.*'"

Perhaps, however, his friend felt lonely sometimes at his house at Selborne, for the letter continues—

"I am glad you have got Mrs. Snooke with you; for I remember that what with snuffing the Candle, making up the wood-fire, and paring your nails, you could seldom get through the writing of one letter in an evening; now she may do a good deal of this for you. But she must not talk; for then you will think of nobody but her; and that must not be. . . . I have but little hope of your thinking much of any particular absent Female, because when you say 'while I, doing no good in my generation, am still single,' you did not insert the lover-like word *alas!* after *I*. There is a sort of sentimental sorrow in the whole sentence, but there is not Feeling enough for a man in earnest without the word *alas.*"

Meanwhile the heart-whole bachelor at Selborne continued to record contentedly in that veritable journal of Adam in Paradise, his *Garden Kalendar*, the exact particulars of brewing his strong beer and bottling "half an Hogsh. of Mrs. Atherley's Port wine," which had, he notices, "not quite so good a smell and flavour as usual, and seemed always to show a disposition to mantle in the glass." Later on in the spring he made his usual journey to Oxford, and paid his brother at Fyfield two visits in the summer. In the intervals he engaged in thinning and tacking his peaches and nectarines "in a very regular manner," and in March sent some verses to Mulso, which the latter pronounced "as good as ever you wrote; they are full of lively Description, of Natural Painting, of Tenderness, and Elegance."

Mulso writes on June 29th, 1764:—

"My brother has described to me your new alterations!"

He goes on to describe a ball given at his house in Yorkshire—

"How would you have stared! and what music book must I have got for you to have studied in a corner?"

Mentioning his having to part with a horse of his own, he remarks—

"I have no idea of a separation between you and a horse, who were once the *centaur not fabulous*."

In September the *Garden Kalendar* records:—

"Got a stone mason to fix the stone with my name and the date of the wall in the middle of the fruit wall. When the mason came to chizzel a hole for the stone he found the wall perfectly sound, dry, and hard."

Part of this wall stands to this day. The inscription is "G. W., 1761." Contented though he seems to have been in his Eveless Eden, his thoughts must have recurred to the subject of matrimony; for on September 14th, 1764, his friend Mulso writes:—

"I find by your hints that you are determined not to die an old Batchelour, if you can help it; tho' you are not yet fixed. Well, speed you well! but look well about you; consider you are beyond your *octavum lustrum,* which, tho' it is (*I hope*) not quite time to leave off the Ladies, is full time to begin with them."

The obstacle to matrimony was clearly the fact that unless he accepted a college living, his fortune, without his Fellowship, which would of course have been vacated by marriage, was insufficient to support a family; and to accept preferment would have probably involved leaving his home and all its belongings at Selborne. Just at this time (at the end of 1764) the college living of Cholderton, in Wilts, on the borders of Hants, fell vacant. In January, 1765, Mulso writes:—

"You have been upon the ramble a good while, and I have had no precise account yet of your Return to Selborne; but from the season of the year I guess that you are at home. To say truth, from a hint in your last letter, and your patient acceptance of *my* hint about matrimony, I suspected that your journey to London had a view to that change of your condition; especially as you show an inclination to detach yourself from College by accepting of so moderate a living as Cholderton. . . . If you can be dispensed with for residing at Cholderton, any little thing added to your own

Fortune would make you comfortable. Wilts is under Salisbury, and the present bishop is a reasonable man. I hear heavy complaints of my Lord of Exeter for obliging his clergy to reside, whether they have a house or no, so I am glad it is not in that Diocese. I had a letter lately from my sister Chapone in which she tells me that she had laid a little plan with you for a visit to me this spring or summer."

Whether the Bishop of Salisbury proved what Mulso would have termed an unreasonable man does not appear, but the living was declined.

On March 9th, 1765, Mulso again writes :—

"I hope to see you according to the plan you have laid down. I have had a letter from Mrs. Chapone, in which she seems to think, that, if business will permit, she cannot have an opportunity of a fellow traveller more agreeable than yourself, and that at her time of life she may set off with you without the imputation of being driving away to Scotland. . . . Your description of your sitting in your dining-room reminds me of your old situation in the Fens. For shame, Gil, this vacuity ought to have been filled up."

On May 19th, 1765, another letter announces the recent dangerous illness of Mrs. Mulso, and that Mrs. Chapone would "not be of your party. She cannot leave London till July, if then, and has another plan of conveyance."

The *Garden Kalendar* pleasantly records its compiler's observations on the journey home from Fyfield, where he had been visiting his brother Henry.

"July 6. There have been fine rains round Andover and Salisbury: the verdure on the Downs is very delicate and the sheep ponds are full of water. But when I came on this side Alresford I found all the ponds without one drop of water; and the turf and corn burnt up in a very deplorable manner; and everything perishing in the gardens. . . .

"The downs between Alresford and Andover are full of Burnet; so full in many places that it is almost the only herb that covers the ground; and is eaten down very close by sheep, who are fond of it. The case is the same between Andover and Sarum where in many places the ground is covered with Burnet now in seed: a child might in those places gather a considerable quantity in a day. It is worth observation that this herb seems to abound most in poorest and shallowest chalky soil. On Selborne Common (a rich strong piece of ground) it has not been yet discovered.

"Near Walker's Ash I rode through a piece of ground of about 400 acres which had been lately pared by a breast plough for burning: here the Burnet was coming up very thick on the bare ground, though the crown of the root must have been cut off of course along with the turf: this shows that it is a plant tenacious of life, since it springs from the severed root like plantain."

On June 24th, 1765, Mulso intimates that he is

"pretty confident that if poor Ned Acton gets better and can spare his assistant, or if any other hand can be got to be employed in your musical affair, that you will set off for Thornhill in a moment, in spite of the new fronting of your stables. You build like rich men, who generally take care of their horses' conveniencies before the rest of their family. . . . To whet your inclinations to come northward, I let you know that my sister Chapone is now with us, and will, I hope, stay with us till the winter.* . . . My sister kindly

* It was on the occasion of this visit to her brother, or perhaps during one paid in 1766, that Mrs. Chapone conceived that partiality for her niece, his eldest daughter, to which society is indebted for her 'Letters on the Improvement of the Mind.'

hastened her coming to be a comfort the sooner to my poor dear wife, who is still in a very weak way. Do you not think it was a bold undertaking in Mrs. Chapone to set out alone from London and be hurried away in chance company in the Leeds machine?"

Probably it was largely owing to his dislike of travelling by the Leeds—or any—machine, that Gilbert White never saw Thornhill, since he suffered severely from "coach-sickness."

The letter continues—

"I saw in my last paper that Dr Blake's death has vacated the living of Tortworth; pray, is not that in your option as Fellow? I am impatient to know whether you are rector of Tortworth; if so, clap in a curate as soon as possible, and come away after institution, induction, etc.; they can better bear your absence before they are well acquainted with you than they can afterwards. Mrs. Mulso and I were very much shocked at the accounts of the accident and end of the poor little boy. It seems, however, to have settled you in your debates upon matrimony, and confirmed you in your state of celibacy; for you observe with a formality of stile, which you drop in the next sentence, that Wedlock *hath* also numbers of cares, etc., as if you had excerpted the observation from a treatise on the expediency of dying an old Batchelor."

Whether he really felt lonely or not, Gilbert White at this time took up seriously the study of botany, purchasing Hudson's 'Flora Anglica' in this year—a book in which he ultimately marked 439 plants as being found in the parish of Selborne.

From August 9th, 1765, the *Garden Kalendar* changes its name to that of a *Calendar of Flora and the Garden.* Fauna, however, was not forgotten—

"July 16. A great rain at Hazelmere where I was then. Several fern owls or goatsuckers flying about in the evening at Blackdown House.

"July 21. The glow-worms no longer shine on the Common: in June they were very frequent. I once saw them twinkle in the South Hams of Devon as late as the middle of September.

"July 28. The goldfinch, yellow hammer, and skylark are the only birds that continue to sing. The redbreast is just beginning. The field-crickets in the Lythe cry no longer.

"August 30. The water wagtail seems to be the smallest English bird that walks with one leg at a time: the rest of that size and under all hop two legs together."

In September her nephew visited Mrs. Snooke at Ringmer, near Lewes. During an eight hours' rain he "lay-by at Brighthelmstone." In a lane "towards the sea, near a village called Whiting," he discovered what he believed to be the Burnet-rose, and notes the wheatears on the Sussex down—

"Vast quantities are caught by the shepherds in the season (about the beginning of Harvest); and yet no numbers are ever seen together, they not being gregarious."

Returning to Selborne he "examined the wild black Hellebore, an uncommon Plant in general, but very common in Selborne Wood."*

* Two species of Hellebore, *fœtidus* and *viridis*, are recorded in his copy of Hudson's 'Flora Anglica,' with the usual + and in addition the mark !

On October 15th he set out for Oxford. On his way he evidently botanised with industry. He notes—

"Oct. 16. Discovered on the banks of the Thames as I walked from Streatly to Wallingford—

"The water hoar-hound, *marrubium aquaticum;* the yellow willow-herb, or loose-strife, *lysimachia;* the purple spiked D° *lysimachia purpurea;* and the Comfrey *symphytum magnum* in bloom, being one of the *herbæ asperifoliæ;* water figwort, *scrop. aquat.*"

The entry continues with the list of plants he noticed at the Physic Garden at Oxford. On returning to Selborne, many botanical discoveries there and in the neighbourhood were recorded.

On April 25th, 1766, John Mulso, who had spent the winter in London, writes expressing his delight at meeting his friend again :—

"Vegetation thrives apace now, and I suppose you are quite intent on your new study. You will not perhaps relish a Prospect the worse when we force you to look up, as I presume you will go with your eyes fixed on the ground most part of the summer. You will pass with country folks as a man always making sermons, while you are only considering a Weed. I thank you for your learned dissertation on the *Canker* or *Stinkpot.* I knew in general that all flesh was grass, but I did not know that grass was flesh before."

Other more practical matters, however, than botany engaged attention at Selborne sometimes; as appears from the following rather curious entry in the *Garden Kalendar*:—

"Ap. 26. Finished moving my barn, which I set at the upper end of the orchard. It began to move on Thursday the 17, and went with great ease by the assistance of about 8 men for that little way that it went in a straight line: but in general it moved in a curve, and was turned once quite round, and half way round again. . . . The barn is 40 feet long."

"June 18. Received a Hogsh. of port from Southton between Mr. Yalden and myself."

On July 3rd, 1766, Mulso recurs to his friend's matrimonial, or perhaps more strictly non-matrimonial, intentions. He was entertaining Mr. and Mrs. Thomas Mulso and Mrs. Chapone at Selborne.

"These summer Visitants of your's are great hindrances to a certain scheme, which you once told me should be soon undertaken, if it ever was to be undertaken at all. It was a scheme to prevent your marrying your maid when you was sixty or seventy. I shall not let you go above a year or two more, before I begin to take the other side of the question, and inveigh against your undertaking this yoke of wedlock at all; and that, because the chances will go more against you, whether you may or may not live to see your children Christianly and virtuously brought up. . . . Come, my dear friend, time is precious. Somehow it has entered into my head that I shall not stay long in this country. Let me have the satisfaction of showing to you a very pretty part of it."

The *Garden Kalendar* continued as usual during this year (1766), but its author composed in addition what he termed—

"FLORA SELBORNIENSIS
with some coincidences of
the coming and departure of birds of passage
and insects; and the appearing of
Reptiles
for the year 1766."

"The Plants are according to Mr. Ray's method; and the Birds according to Mr. Willughby's ornithology; the Insects according to Ray's 'Hist. Insect'; and the Reptiles according to Ray's 'Synopsis Animalium Quadrupedum.'"

The matter of this little work is, both as regards the botany and zoology, of a much more elementary character than was eventually attained to in 'The Natural History of Selborne,' yet it is interesting as showing the first idea of an 'Annus Historico-Naturalis,' which afterwards took shape in its author's mind.

On April 13th, 1767, John Mulso became, by the collation of the Bishop of Winchester, Vicar and Rector of Witney, Oxfordshire. While congratulating his friend that their communications would now be more easy, he regrets his never having seen Thornhill. The letter is signed "affectionately yours (as the Bishop calls me) Witney John."

This summer is remarkable for the commencement of the correspondence with Pennant, the first letter being dated August 10th, 1767.* This letter, with

* The date is given as August 4th in 'The Natural History of Selborne,' for which this, with other letters, was transcribed. For some reason (or perhaps for no reason) the dates of the original letters were changed in many instances by their author when preparing them for publication.

some excisions, became the tenth in 'The Natural History of Selborne,' the first nine letters describing Selborne and its neighbourhood being written subsequently for the book. On the 18th of the preceding April Gilbert White had gone to London, and perhaps he had been then introduced to Pennant by his brother Benjamin, who was Pennant's publisher.

At this time Thomas Pennant, a Flintshire gentleman of good family and fortune, and with literary tastes, was forty-one years of age. In the year 1755 he had commenced a correspondence with Linnæus, and in 1761 had published the first (folio) edition of his 'British Zoology.' On February 25th, 1767, he had been elected a Fellow of the Royal Society, and he mentions in his 'Literary Life,' published in 1793, that Mr. Benjamin White, bookseller, proposed to me the republication of the 'British Zoology,' which was carried into effect in the succeeding year (1768).

The original letter of August 10th, 1767, addressed "— Pennant, Esq., at Downing, in Flintshire, North Wales," commenced—

August 10, 1767.

At Selborne, near Alton, Hants.

Sir,—Nothing but the obliging notice you were so kind as to take of my trifling observations in the natural way, when I was in town in the spring,* and your repeated

* From this sentence one would naturally infer a *personal* introduction to Pennant. Yet this seems to be distinctly negatived by a remark of Gilbert White's in his original letter to Pennant of 28th February, 1769, and by another in the original letter of 30th March, 1771.

mention of me in some late letters to my brother, could have emboldened me to have entered into a correspondence with you: in which though my vanity cannot suggest to me that I shall send you any information worthy your attention, yet the communication of my thoughts to a gentleman so distinguished for these kinds of studies will unavoidably be attended with satisfaction and improvement on my side.

The letter concludes with a line or two concerning the *Falco* which it mentions, and is subscribed—

"I am, with the greatest regard,
Your most humble servant,
GIL. WHITE."

In his edition of 'The Natural History of Selborne'* Mr. Bell has commented in rather severe terms upon the conduct of Pennant, whom he terms "vain and self-seeking," towards his distinguished correspondent; and he is in error when he states that "there is no acknowledgement of his help, no recognition of his debt," since the "Rev. Mr. White of Selborne, Hants," is mentioned among the "learned and ingenious friends" to whom Pennant acknowledges his indebtedness in the Preface to the second (1st 8vo) edition of the 'British Zoology,' published in 1768.

On the whole, though of course there is no comparison between the information afforded by the original and scholarly Naturalist of Selborne, and that of his industrious, but perhaps a little too complacent correspondent, the relations between the

* vol. i. p. lxi. *et seq.*

two seem only to have been pleasant. One little cloud arose, however, as will appear later on, when Pennant seemed to be reluctant to part with certain drawings of some of John White's natural history specimens, intended to illustrate the latter's 'Fauna Calpensis.'

On July 21st, 1767, Mulso exhorted his friend to leave his *Tiburni lucum*, and deign to visit him at Witney. He wrote again on October 13th, with the information that Mr. Frewen, the Rector of Cromhall, was dead.

"This you will have known I presume by a speedier intelligence. But it is not so much the thing itself, as the inferences, that affect me. You are said to be likely to take his living. If you do, two or three things will happen. You will come soon to Oxford, tho' you are not pressed in time. You will keep a curate on that living, and therefore not be so tied by the leg, as you now are by your serving a curacy yourself; a circumstance very hateful to a man whose inquisitive genius makes him love to change the scene often and search for curiosities in various regions. This Living is in Gloucestershire, as I understand: I lye in the very road, so that upon the whole, I conclude from these inferences, I may see you soon, and I may see you often. I shall have you routed out of that Recess of Selbourne, where your affections are too much engrossed for yourself, and your friends at a distance.

"I am afraid that this is not the best living of the College; but nevertheless I think I collected by our last confabulation, that you was inclined to secure to yourself the first thing that fell, and get rid of your Fellowship before your Fellowship got rid of you."

The writer rightly concluded that his friend would come and visit him at Witney, whence he "rode out on purpose to look after the base hore-hound, the *stachys fuchsii* of Ray, which that gent. says grows near Witney Park." Possibly he was more interested in botany than in the advantages of the Rectory of Cromhall, since he did not accept that preferment. At this time he made entries in the *Garden Kalendar* of a point in natural history which never escaped his attention, and to which he constantly recurred—the late appearance of swallows.

"Oct. 29. Saw four or five swallows flying round and settling on the County Hospital at Oxoñ."

"Nov. 12. Bro. Benjn saw a Marten flying in Fleet Street."

In December, 1767, a very interesting entry occurs respecting the harvest mouse, which Gilbert White was the first to discover and describe in this country.

"Dec. 4. Sent two field mice, a species very common in these parts (tho' unknown to the Zoologists) to Thomas Pennant, Esq. of Downing in Flintshire. They resemble much in colour y^{e} *mus domesticus medius* of Ray; but are smaller than the *mus domesticus vulg. seu minor* of the same great Naturalist. They never enter houses; are carryed into ricks and barns with ye sheaves; abound in harvest; and build their nests, composed of the blades of corn, up from the ground among the standing wheat; and sometimes in thistles. They breed as many as eight young at one time."

Pennant availed himself readily of his correspondent's information. He has been blamed by

Mr. Bell for not giving his friend the credit of the discovery; but he did attribute it to Gilbert White in the second edition of his 'British Zoology.'

At the commencement of 1768, the *Garden Kalendar*, which up to this date had been written on quarto letter-paper, changes its form; being replaced by an oblong book, *The Naturalist's Journal*, which was, its owner records therein, "The gift of the Honourable Mr. Barrington, the Inventer." Benjamin White was the publisher.

The pages of this book have printed headlines over ruled spaces. Forms have, indeed, some advantages; but they are certainly liable to too much elaboration, being, like some other things, good servants but bad masters. It is noticeable that at first the compiler was scrupulous to keep exactly within the boundaries of the dividing lines; but gradually, and to the distinct advantage of his records, he overstepped them, and gave the most space to the matters that most engaged his attention; and as years went on even used blank leaves for any special matter. The records continued, as in former years, to be of the most varied description. In addition to natural history observations (of which those on insects become more frequent), containing in the rough very many of the facts from which he composed his book, he notes family events, visits to and from his friends, his garden produce, the appearance of his neighbours' crops and their methods of cultivation. In short, the book continues to deserve the name of 'The Journal of Adam in Paradise.'

On January 30th, 1768, a report of the death of the Provost of Oriel caused Mulso to write in haste with the news to his friend, in the hope that at least it might bring him up to Oxford and his neighbourhood. The report proved true, Provost Musgrave having been found dead in his room on January 29th.

Gilbert White went to Oxford from London, where, as usual at this time of year, he was visiting his brothers; but it does not appear that he stood again for the Provostship, to which Dr. John Clarke was elected on February 12th, 1769.

The commencement of the original letter to Pennant* of March 30th, 1768, which became Letter XV., contains an invitation to that gentleman, who was contemplating a visit to see the moose deer at Goodwood, to visit him at Selborne, and offering facilities for this. The visit was not, however, made.

The letter ended—

"When your sheets containing a list of the British birds etc. come out, you will gratify me much by your sending me one. I am glad to hear you intend to continue your publications in the natural way. My relation at Gibraltar had never at all applyed to these kind of studies, and has no books of that sort: else he might be helpful to you with

* The more interesting unpublished passages of these letters are now printed.

regard to the birds of Barbary and Andalusia. Pray give my humble respects to Mr. Banks and tell him I shall not forget him next month with regard to the *Lathræa Squammaria.* If he will do me the honour to come and see me he will find how many curious plants I am acquainted with in my own Country. I request also that you will be pleased to pay my compliments and thanks to Mr. Barrington for the agreeable present of his Journal, which I am filling up day by day. Buntings I saw in plenty last week."

Letter XVI., written on April 18th, 1768, commenced—

"As I had set my mind on the pleasure of your conversation, so I was in proportion disappointed when I found that you could not come. But as your business may be over now, I shall still live in hopes of seeing you at this beautiful season, when every hedge and field abounds with matter of entertainment for the curious. If you could come down at the end of this week, or the beginning of next, I should be ready to partake with you in a post-chaise back to town on the second of May."

On June 2nd, Mulso writes :—

"I hope Mr. Etty rejoices in his agreeable accession of preferment; I wish you had obtained your views in that neighbourhood."

Mr. Etty's new living was the vicarage of Whitchurch, near Pangbourne, to which he was presented by Lord Chancellor Camden. Mulso's hint as to his friend's views referred, as appears later on, to the Oriel living of Ufton Nervett in Berkshire. In July Gilbert White visited his brother Harry at Fyfield, whose rectory there was the "Gentleman's house," where, he tells Pennant (in Letter XVIII.), he had neither books on natural history, nor leisure to write.

About this time the living of Cholderton, Wilts, which had already been refused, became again vacant.

Mulso writes on July 26th, 1768 :—

"I long to know your determination about Cholderton; Dr. Bentham was of opinion that you would take it. He talked something of the vicinity of it to Harry, as if he might avail himself of your curacy, or be at hand to take the Parish duties. How comes it to pass that you, who want to make Selborne your residence, are afraid of a living where your residence would not be required? This is one of those Paradoxes in which you have always delighted. For it does not follow that you may not visit your Parishioners very often, though you do not inhabit the village; and that is more than you did at Moreton Pinkney. If you once make your living your residence, farewell poor Selborne! Not that I am at all solicitous about your taking the present thing, if you have a good prospect of the *Principal Object** of your expectations: it is certain that you can very well wait, if it would tally at last, and be a gainer instead of a loser."

He goes on to say—

"Bon voyage to Mr. Banks! What a fund of entertainments will he have laid in for you, if he lives to come back!"

The nature of Gilbert White's acquaintanceship with Mr. (afterwards Sir Joseph) Banks does not exactly appear. Writing to Pennant, however, on June 10th, 1768, after mentioning that while in town he "was often in company with your friend Mr. Barrington," he continues—

* Ufton Nervett Rectory.

"Even Mr. Banks (notwithstanding he was so soon to leave the kingdom and undertake his immense voyage) afforded me some hours of his conversation at his new house, where I met Dr. Solander."

Again, in another letter to Pennant under date October 8th, 1768*—

"I met with a paragraph in the newspapers some weeks ago that gave me some odd sensations, a kind of mixture of pleasure and pain at the same time. It was as follows: 'On the 6th day of August, Joseph Banks Esq., accompanyed by Dr. Solander, Mr. Green the astronomer, &c., set out for Deal in order to embark aboard the 'Endeavour,' Captain Cook, bound for the South Seas.'

"When I reflect on the youth and affluence of this enterprizing gentleman I am filled with wonder to see how conspicuously the contempt of dangers, and the love of excelling in his favourite studies stand forth in his character. And yet though I admire his resolution, which scorns to stoop to any difficulties; I cannot divest myself of some degree of solicitude for his person. The circumnavigation of the globe is an undertaking that must shock the constitution of a person inured to a sea-faring life from his childhood; and how much more that of a landman! May we not hope that this strong impulse, which urges forward this distinguished naturalist to brave the intemperance of every climate, may also lead him to the discovery of something highly beneficial to mankind? If he survives, with what delight shall we peruse his Journals, his Fauna, his Flora! If he falls by the way, I shall revere his fortitude, and contempt of pleasures, and indulgences: but shall always regret him, though my knowledge of his worth was of late date, and my acquaintanceship with him but slender."

* These passages occur in the original letters to Pennant, which became Letters XVII. and XX. of 'The Natural History of Selborne.'

In November of the same year he again writes :—

"Poor Mr. Banks! his undertakings are virtu in excess; and I could almost wish he had followed your advice, and sent a proxy.* But then he would have foregone the honour and praise due to such a disinterested hazarding of his life; which a very sensible man the other day told me much more merited a peerage than the enterprise undertaken by Lord Anson."

On December 8th, 1769, writing to Pennant, he refers to the difficulties which the expedition had experienced at the hands of the Portuguese Government :—

"It is grievous to see from Dr. Solander's letter in the 'Gentleman's Magazine' dated from Rio de Janeiro with what insolence the Viceroy of Brazil treated those gentlemen who have hazarded their lives in the pursuit of natural knowledge; and this is not the worst of it: for when they arrive in the South Seas their reception will be just the same from every Spanish Governor from Chile to Mexico."

Mr. Banks' safe return in 1771 is referred to in letters to the same correspondent of July and September of that year.

A short letter to Mr. Banks from Gilbert White, dated April 21st 1768, is printed by Mr. Bell,† in which he regrets that his correspondent and Mr. Pennant could not visit him at that time, "but now it seems that I must plod on by myself with few

* Pennant paid the Rev. George Low to explore Orkney and Shetland for him in 1774 and 1778.

† *Vide* Bell's edition of 'The Natural History of Selborne,' vol. ii. pp. 241–242.

books and no soul to communicate my doubts and discoveries to." He concludes by wishing Mr. Banks "all health and a great deal of success and satisfaction in your laudable pursuits, a prosperous voyage and a safe return."

Returning to the unpublished passages from the original letter to Pennant; Letter XVII of June 18th (10th in the original letter), 1768, continued—

"I am now to return you my warmest thanks for your agreeable present of the 'British Zoology,'* which I accept with great satisfaction as a token of your friendship: and shall look upon your work as an ornament to my little shelf of natural history. As far as I have been able to compare any animals with your descriptions, I find them just and apt; and such as may readily help the reader to ascertain any quadrupede or bird that falls in his way."

On August 17th, 1768, what became Letter

* This was the *second* edition, of which, at Benjamin White's proposal, Pennant (see his 'Literary Life,' p. 8) this year published with him two volumes containing the quadrupeds and birds—a third volume, including the reptiles and fishes, appearing in 1769. In the Preface the author gratefully acknowledges the information received among others from Gilbert White, and in the Appendix (p. 498), which must have been written in 1767, gives an account, almost in the latter's words, of the "non-descript species [of field-mouse] communicated to us this year by the rev. Mr. *White* of *Selborne, Hants*," besides mentioning him (p. 500) in connection with the tame bat and the falcon, as related by him in his letters of August-November, 1767. A fourth and supplementary volume (often called the "third edition") was published three years later (see note, p. 176, to letter of 12th May, 1770). Herein Pennant again refers to Gilbert White giving his observations on migrating swallows seen on Michaelmas Day, 1768.

The key to nearly all the subsequent letters to Pennant is that White was continually furnishing information additional to that contained in the second edition of the 'British Zoology.' This information was eventually incorporated in the "*fourth*" edition, published in two forms (4to and 8vo) in 1776, of which Gilbert White, while in London, himself corrected the proofs (see note to letter to John White of 26th February of that year).—A. N.

XIX. was addressed from Selborne to Pennant. The letter originally contained the following passages :—

"I wrote to you about the 25th of July, and hope my letter reached you, as it was directed to Sir R[oger] Mostyn as usual. In that letter I gave you an account that I had sent the fishes of our streams up to Mazel to be engraved. You had in it also a pretty exact description of the Ambresbury loach taken from living specimens procured from thence; my sentiments with regard to the use of toads near Hungerford; and my suspicions with regard to the water-eft."

[Here follows the paragraph already printed at page 115.]

"The behaviour of the antelope which you saw in town, strongly corroborates my suspicions concerning the deer. I desire you will not fail to procure a buck's head from Sir R. M. and will have it dissected with care. I could have procured one with ease myself 'til this year: but now my neighbour Sir Simeon Stuart has destroyed his stock, and turned his park into sheep-walks.

"The first young swallows appeared on July 4th, and the first martins began to congregate on the bush of the village maypole on July 23.

"To me it is very plain that the first swallows and martins that congregate, are the birds of the first brood, and that thro' an inability of flying long at a time. For while these first flyers are spending as much time on a maypole, the battlements of a tower &c.: the old ones are busily employed in rearing a second brood.

"The swifts have never been seen with us since Aug. 5th, and I conclude will come no more this season. I am always amazed that this species should constantly depart so many months before their congeners. It is worth our remarking that tho' the swift is at its arrival of a dark sooty colour; yet by being for 16 hours together almost constantly in the sun and air, it becomes before its departure much bleached, and as it were what the country people call piss-burnt, like an old weather-beaten brown wig: and yet it returns sooty again in the Spring. Now if they go into warm regions during our winter, why do they not return sun-burnt, as they went off? It is a matter of curious enquiry to consider when swifts moult. Change their feathers here they certainly do not: and if they have as much occasion for their wings while absent, as while here, they would find no opportunity to spare several feathers at a time. I would not pretend to lay too much stress on these reflections, but certainly can't refrain from observing that they tend rather to make one suspect that they hide, rather than migrate—at least for *part* of the long time they are absent from us; and perhaps at that juncture they moult."

Letter XX. to Pennant, dated October 8th, 1768, began—

"Your letter of September 6th gave me a great deal of entertainment and satisfaction; and the more satisfaction because I really began to fear from your long and unusual silence that you might be prevented from writing by sickness or some accident that might have befallen you in your Caernarvon tour. But as I much esteem your friendly correspondence already, I desire you will not make use of any such methods of enhancing the value of it for the future.

"I receive your kind invitation into Flintshire as a fresh instance of your friendly disposition towards me; but whether my health, or the want of command of my time will ever permit me to gratify myself with so pleasing a tour and visit, I cannot pretend to say; however I depend much on having it in my power to give you a meeting in town next spring: and it would be a matter of high entertainment and instruction to me to be able to accompany you in your pursuits after natural knowledge."

Letter XXI. to Pennant, written at this time, November 28th, 1768, originally commenced—

"Your obliging and communicative letter of October 23rd lies before me; and ought not any longer to remain unanswered. It is a great pleasure to me to find that amidst your various and extensive correspondence, and the daily labours of your work in hand, you still afford time to pay regard to my trifling remarks, and discoveries; which a man cannot avoid stumbling upon now and then, if he lives altogether in the country, and gives any attention at all to the works of Nature. Happy the man! who knows, like you, how to keep himself innocently and usefully employed; especially where his studies tend to the advancement of knowledge, and the benefit of Society. And happy would it be for many more men of fortune if they knew what to do with their time; if they knew how to shun 'the pains and penalties of Idleness,' how much dissipation, riot, and excess would they escape; not without the complacency of finding themselves growing still better neighbours, and better commonwealths-men?"

Further on the following passage occurs:—

"I have also written to my South country correspondent in Gibraltar, and urged him to take up the study of Nature a little; and to habituate his mind to attend to the migrations of birds and fishes; and to the plants, fossils, and insects of that part of the world. I have also sent him your 'British Zoology' that he may see what is going on at home: and my Brother has sent him Ray's *Synopsis avium & piscium*, the *Systema Naturæ*, Ray's *Synop. animalium quadrup*m &c. As to birds I fear that the concourse and din of a garrison will not prove very inviting to such timid animals: and long or frequent excursions into Andalusia may not be allowed of by the bigotted and narrow-minded Spaniards; nor be consistent with the strict and rigid discipline of a place at arms surrounded with a constant blockade of jealous enemies. However I could earnestly wish to see a well executed Fauna from that part of the world.

"It is a matter of no small satisfaction to me to hear that you are so forward in your work, and that it is to appear in the spring."

Letter XXII. to Pennant, dated January 2nd, 1769, contained originally the following passages:—

"In your letter of June 28th, 1768, I could but admire with how much frankness you acknowledged several mistakes in your 'Zoology' with respect to some birds of the *Grallæ* order. Candor is a very essential part of a Naturalist, and this accomplishment our great countryman Mr. Ray possessed in an eminent degree; and that rendered him so excellent.—If a man was never to write on natural knowledge 'til he knew everything, he would never write at all: and therefore a readiness to acknowledge mistakes on due conviction is the only certain path to perfection. . . . I often take up your 'Zoology' for an hour, and entertain myself with comparing your descriptions with those of the authors that have written on the same subject; and am pleased to find that my friend has thro' the whole acquitted himself so much to advantage. Your treatise in particular on migration I admire much, and think that if it is enlarged as more information comes in, it will contribute much to the advancement of natural knowledge. But there is a passage in the article Goatsucker, page 247, which you will pardon me for objecting to, as I always thought it exceptionable: and that is 'This noise being made *only* in its flight, we suppose it to be caused by the resistance to the air against the hollow of its vastly extended mouth and throat for it flies with both open to take its prey.' Now as the first line appears to me to be a false fact; the supposition of course also falls to the ground, if it should prove so."

The passage in question was omitted in subsequent editions.

Writing to his friend, who was visiting his brother at Fyfield, on January 28th, 1769, Mulso says:—

"I am glad to hear that Harry enjoys the blessing of increasing and multiplying: for tho' it makes a house strait, it makes it chearful. This you know tho' still a Batchelour; for no man is more free to fill his house, than yourself. . . . I am not at all surprized at the pleasure you take in your pursuits of natural knowledge. I know nothing more capable of satisfying the curiosity of the human mind which is always searching after novelty, for the subject is inexhaustible. . . . I shall endeavour to remember to say *Laurustinus* hereafter, and you may farther inform me how I may with propriety use it in the plural number."

The original letter to Mr. Pennant, which was written at this time, February 28th, 1769, and became Letter XXIII. of 'The Natural History of Selborne,' commenced with these paragraphs:—

"Dear Sir,—Some avocation or business of one kind or another has still prevented my paying that attention to your kind letter of Jan. 22: which it deserved. As at the close of that letter you invite me in a most obliging manner to come and spend some time in Flintshire; that paragraph seems to challenge my first attention. You will not, I hope, suspect me of flattery when I assure you that there is no man in the kingdom whom I should visit with more satisfaction. For as our studies turn the same way, and we have been so well acquainted by a long and communicative correspondence, I trust we should relish each others conversation, and be soon as well acquainted in person as by letter. Besides your part of the world would not be without its charms from novelty; as I am not acquainted with the N.W. part of this island any farther up than Shrewsbury. Your improvements, your mines, your fossils, your botany, your shores, your birds, would all be matter of the highest entertainment to me. But then how am I to get at all these pleasures and amusements? I have neither time nor bodily abilities adequate to so long a journey. And if I had time I am subject to such horrible coach-sickness, that I should be near dead long before I got to Chester. These difficulties, I know, will be matter of great mirth to you, who have travelled all over Europe; but they are formidable to me. As therefore the man cannot come to the mountain, I hope the mountain (since friendship will effect strange things) will come to the man: I hope you will have it in your power to meet me in London, and that you will gratify me with an opportunity of waiting on you to Selborne."

In April a visit was paid to London, where amusement was found for several days in observing and noting all the different fish in the market.

A letter to Pennant of May 29th, 1769, which became Letter XXIV., commenced as follows:—

"Dear Sir,—When your agreable but tardy letter of April 22nd arrived at this place, I was in London: but it was sent up after me. It gave me concern to hear you had been a good while indisposed; and satisfaction to find that you are recovered.

"The great honors that have befallen you at Drontheim call for my congratulations.* You must heartily believe now

* Writing of 1769 Pennant says ('Literary Life,' p. 11): "In the same year I received a very polite letter from the reverend *Jo. Ernest Gunner*, bishop of *Drontheim* in *Norway*, informing me that I had been elected member of the Royal Academy of Sciences on *March* the 9th past; of which

FLEET STREET IN THE EIGHTEENTH CENTURY

[To face p. 168, Vol. I.

in the accounts given by Pontoppidan of the kraken, and sea-snake: if you should express any disrespect towards these two remarkable animals, I don't know but they may remove you from y[e] society as an unworthy Brother."

On June 30th of this year, 1769, the first letter to the Hon. Daines Barrington, to whom no less than sixty-six letters were published as addressed, was written by Gilbert White.

This gentleman, who has been described as "a queer compound of the lawyer, antiquary, and naturalist," was son of the first Viscount Barrington. Born in 1727, he went to Oxford, was called to the Bar, and became a Welsh Judge, and subsequently, in 1764, Recorder of Bristol. He died in 1800. He was a Fellow of the Royal Society, and it was through him that Gilbert White's account of the *Hirundines* was read before that society in February, 1774, and March, 1775.

The personal acquaintance with Barrington did not begin till May, 1769, when they met in London, though more than a year before, as already stated, Gilbert White received from him the printed *Naturalist's Journal;* for an unpublished portion of Letter XIII. to Pennant, dated January 22nd, 1768, mentions—

society that prelate was president." It would seem that Pennant lost little time in communicating this fact, of which he had every reason to be proud, to his correspondent, and the humorous turn the latter gives to the announcement is eminently characteristic of him. The marvellous account of sea-monsters given by Pontoppidan, who was Bishop of Bergen, in his Natural History of Norway (vol. ii. pp. 297–354), was published at Copenhagen in 1753, and, through the English translation of 1755, has long been notorious.—A. N.

"Your friend Mr. Barrington (to whom I am an entire stranger) has been so obliging as to make me a present of one of his Naturalist's Journals, which I shall hope to fill in the course of the year."

It was to Pennant, then, that the introduction to Daines Barrington was probably owing,

Barrington gives honourable mention to his friend and correspondent in his 'Miscellanies,' published in 1781, as a "well-read and observant naturalist." And Gilbert White returned the compliment in Letter LI. to him, in which he writes, "I have now read your 'Miscellanies' through with much care and satisfaction."

After a visit to Oxford in July, 1769, when he noticed "vast flocks of young wagtails on the banks of the Cherwell," the Naturalist received his friends, Mr. Skinner, of C.C.C., Oxford, and Mr. Sheffield, of Worcester College, at Selborne, the latter of whom, as he records, "went into Wolmer Forest and procured me a green sandpiper." Of this visit Mulso remarks in August that "the world will, I presume, be the better hereafter for your joint labours."

The following formed the commencement of what became, when revised, Letter XXV., September 1st, 1769, to Pennant:—

"I am to acknowledge my tardiness in answering your kind Letter of June 9th, and have to plead business, workmen, and company; and yet I ought not to have been silent for so many weeks.

"In a former letter of May the 9th you mention a

thought of a periodical publication, that shall receive the various pieces of natural history that otherwise might perish. Not being conversant in such undertakings, I am little of a judge whether such a pamphlet would be likely to take; and am fearful that the very occasion of your magazine may be the cause of its not succeeding: for amidst the din and clamour of party Rage, the still small voice of Philosophy will, I fear, be little attended to. However, if you think such a publication expedient, you will no doubt get considerable assistance from your friends; and I shall be ready to advance my mite: but then I shall expect you to be very charitable in your allowance, and to grant that my mite in one respect is equal to larger contributions, as it is all my stock of knowledge."

The letter contained also the following passages :—

"When an opportunity occurs I shall be glad to look into your 'Indian Zoology.' Mr. Skinner of C.C.C. and Mr. Sheffield of Worcester College have lately been with me for a fortnight, and are the only Naturalists that I have ever yet had the pleasure of seeing at my house. They are both excellent botanists, and the latter makes a very rapid progress in Entomology. There was great satisfaction in walking out with these men: because no bird, plant, or insect came before them unascertain'd. One day we shot a *Tringa ochrophus*,* which is a very rare bird in these parts. . . .

"There appears a Comet nightly (having a tail about six degrees in length) in the constellation of Aries, between the 24: 29: and 51 stars of that constellation in the English catalogue."

On October 28th, 1769, the *Naturalist's Journal* contains the entry, "Mrs. J. W. sailed" on her return

* The green sandpiper above mentioned.—A. N.

to Gibraltar. This was his brother John's wife. She had brought to England her only child, who afterwards became known in the family as "Gibraltar Jack."

His thoughts were evidently turned to his brother's collections at Gibraltar, for the *Naturalist's Journal* now contains "a proper antiseptic substance for the preservation of birds, etc.," together with directions for applying it, and also a quotation from Linnæus' 'Amœnit. Academ.' vol. iv., as "a motto for my brother John's Naturalist's Journal kept at Gibraltar." And again, "my brother John's birds are preserved with salt, allom, and pepper."

It had been a disappointment that an intended visit to Selborne by John Mulso and his family had been given up. On November 1st, 1769, his friend was addressed—

"Dear Gil,—You are a man, as I have long known, so very much master of your Passions, and so guarded in your behaviour and even in your expressions, that when I see a little ebullition I guess there is a considerable fire beneath. Thus in your letter that lies before me I see that you was more than commonly disappointed in not having my wife's and daughter's Company and mine at Selborne this summer. . . . You hardly now know what you ask, when you ask for our Company, and for a *good while.* We are expensive and we are troublesome guests. We both cry out '*non sum qualis eram.*' . . . We are greasy, sedentary, potatious Inmates. Take us therefore if you dare, but let there be some agreeable female Companion at hand to sit with Mrs. Mulso in the Bottom, while I once more wheeze and sweat to arrive at the Top of the charming Hanger. My Jenny will be an almost indefatigable Companion, and an assiduous scholar in your botanic searches: you have already taught her to be peering at the bottoms of old walls, tho' a S-r R-v-rence lies close to her Foot. I am glad to hear that your brother John has made so handsome a contribution to your *Feet Measure* of nephews and nieces, and that you all approve of the mother of the boy. I wish him success, and do not wonder that he is tired of the Rock, but yet I think that he is more like to lay up there, than in any place that he can change for in England. However, he is an *Emeritus* and has deserved indulgence. If he proceeds as a southern naturalist, he ought to be under the pay of his brother Ben; for he has had fine pickings out of your naturals—I mean your naturalists—of late years."

The original of Letter XXVII. to Pennant, written from Selborne on February 22nd, 1770, commenced as follows:—

"Dear Sir,—In the first place I am to acknowledge your favour of Decem^r 23: which I had no proper leisure nor opportunity of answering before the time at which you proposed to leave Flintshire. I am also to express my thanks for your friendly letter of last week from London, in which you press me to give you a meeting in town. If nothing was wanting but inclination I should with pleasure have set out before now: but this is not a convenient season for me to be from home; and I am now become a very bad traveller: however, I will endeavour to give you a meeting if possible. . . ."

The letter, as printed by its author, concluded with the sentence—

"In general, foreign animals fall seldom in my way; my little intelligence is confined to the narrow sphere of my own observations at home."

In the original there followed—

"As a naturalist I may say—

'. . . . Ego apis matinæ
More modoque
Grata carpentis thyma per laborem
Plurimum, circa nemus uvidique
Tiburis ripas, operosa parvus
. . . fingo'"—

a favourite quotation from Horace, which is sometimes prefixed to the year's records in the *Naturalist's Journal.* Mulso writes again on March 18th, 1770 :—

"Your time is near to have a call to Oriel, and I hope you will take some time for a visit at Witney. You may there enliven my hours, and enlarge my ideas; and you who *ascertain* everything may ascertain my health, for I am but as grass and as the flower of the field. . . . The secret is out with regard to the *old man of the hill.* My wife thought it improbable that it ever would be brought to bear that she [Jenny Mulso] should see Harry at Selborne, and therefore she explained the matter. . . . I was sorry that my wife judged it best to put a stop to her inquisitiveness, for it was innocent and often ingenious; but her speaking of this to others, and our being obliged to disguise the truth before her, gave us the air of Romancers. However, this has not taken off her desire of seeing your Retreat."

This last passage settles the question as to the identity of the personator of the Hermit at Selborne.

Mulso continues—

"(If you can let me know that a vacancy is likely to happen in the Stalls of Winton, it would be good news; for my uncle having lately provided for Mr. Rennell,* I have hope of being next in succession; but this is *sub sigillo.*)"

* His domestic chaplain.

CHAPTER IX.

In February, 1770, Gilbert White, who was visiting his brother at Fyfield, went to "Charlton in Wilts," and notes in his *Naturalist's Journal* that he

"saw Bustards on Salisbury Plain. They much resemble fallow deer at a distance. Partridges pair, Wild Geese in the winter do great damage to the green wheat on Salisbury Plain. Grey crows are not seen 'til we come about Andover from the eastward. As you go thence westward into Wilts they abound. Buntings abound in this part of Wilts."

This spring apparently marks the first record of intention to collect and publish his observations in natural history, since, on April 12th, just after making his usual Oxford journey, he wrote to Daines Barrington :—

"When we meet, I shall be glad to have some conversation with you concerning the proposal you make of my drawing up an account of the animals in this neighbourhood. Your partiality towards my small abilities persuades you, I fear, that I am able to do more than is in my power: for it is no small undertaking for a man unsupported and alone to begin a natural history from his own autopsia!

Though there is endless room for observation in the field of nature which is boundless, yet investigation (where a man endeavours to be sure of his facts) can make but slow progress; and all that one could collect in many years would go into a very narrow compass."*

Letter XXVIII. to Pennant, dated in the original May 12th, 1770, thus commenced—

"Dear Sir,—A journey of business, which detained me longer from home than I expected, must be my excuse for neglecting to answer your letter 'til this time.

"My thanks are due for your obliging present of your last publication,† which will conduce much to illustrate, and improve the 'British Zoology': the designs are just and the attitudes easy and natural; and the plates so well engraved, that they will convey a much more adequate Idea of an unknown animal to a young naturalist than words possibly can.

"Tho' you are embarked in a more extensive plan of natural history, yet I am glad to find that you do by no means give up the 'Brit.[ish] Zoology'; that I think should be your principal object; and I hope you will continue to revise it at your leisure, and to retouch it over 'til you have rendered it as perfect as the nature of the work will admit of. If people that live in the country

* *Vide* 'The Natural History of Selborne,' Letter V. to Barrington.

† This must have been the 103 plates and 96 pages of text, which were issued in this year by Pennant, forming (as already mentioned p. 162, in the note to letter of 10th June, 1768) what has been called the "*third* edition" of the '*British* Zoology'; but being in reality only the fourth and supplementary volume of the *second* edition. It was printed at Chester by Eliz. Adams, who also printed the third volume of the former edition; but, unlike that, does not bear Benjamin White's name, as publisher, in the title-page. Complete copies of this volume are by no means common. The plates well deserve White's praise of them. They are not signed, but were presumably engraved by Mazel.—A. N.

would take a little pains, daily observations might be made with respect to animals, and particularly regarding their life and conversation, their actions and œconomy, which are the life and soul of natural history."

From the period of Gilbert White's life now reached there is extant a considerable number of his letters, chiefly to members of his family. Most of them are here given; since, especially in the case of such an uneventful life, there can be no better illustrations of their author's thoughts and character. Moreover, now that all those who knew him have long passed away, there is really no other trustworthy means of illustrating his career. These letters were of course never intended for publication, nevertheless many of them have no little literary and scientific interest; while all of them exhibit their author as a kindly, affectionate, and high-minded man.

It should perhaps be here mentioned that many of the letters in question—but by no means all—were lent to Mr. Bell. These he printed in what—originally intended to be an appendix—became a second volume of his edition of Gilbert White's book. In some cases, unfortunately, he appended wrong dates, and even mistook the names of the persons to whom the letters were addressed. Moreover some omissions occur not infrequently, and there are errors in the text.

The following letter is printed from MS. in Gilbert

White's handwriting. It is unsigned, and is probably a copy kept for the purpose of future reference.

*To the Rev. John White.**

[Headed "Bro. John."]

(I.)

Selborne, May 26, 1770.

Dear Brother,—I am to acknowledge first the receipt of your kind letter of Feb. 19, which I should have answered before now, had I not waited for your box of curiosities; concerning which you would naturally expect I should give some account. Farther obligations are now due for a second letter of April 14; but tho' I have not yet received the box, I must no longer omit to take notice of your agreeable communications. It is probable the box may be in London; but I have lately intimated that I would not wish to have it sent down at present, as I hope to be in town as soon as Whitsontide is over.

Your *Vespae* with purple wings are a beautiful and scarce species: they are the *Vespæ crabroni congeneres in Italia captæ* of Mr. Willughby, well described in Ray's 'Hist. Insectorum,' p. 250. Pray observe what they feed on; and enquire into their manner of nidification. Your butterfly-like insect with long remiform wings is curious and rare, and proves to be the *Panorpa coa* Lin. You see it is to be found in few places; and Scopoli† knows nothing of it, though Carniola lies in a warm latitude. Send some more specimens. Pray observe how and where they breed. I suspect much that they come from the

* At Gibraltar.

† This refers to Scopoli's 'Entomologia Carniolica,' published at Vienna in 1763, and not to his 'Annus Primus Historico-Naturalis,' to which reference is often made by Gilbert White.—A. N.

water, where they perhaps are hatched like the *Ephemeræ* (may-flies) and the *Phryganeæ* (cadews).

Here it will be proper to remark that Lin. is too general in some of his assertions: too many exceptions occur under his general rules: as you must have already observed in the course of your reading the 'Syst. Nat.'

You will be pleased to observe whether your ant-catching *Sphex* (for a *Sphex* I certainly think it was, though we soon lost our single small specimen) does not carry its prey to its nest in order to feed its maggots: in and with what substance does it make its nest? I have named it *Sphex formicarum falco.* The Insect with a long slender petiolus between the thorax and abdomen is a fine sort of *Ichneumon.*

Look after the genus of birds called Petrels; they are very peculiar in their way of life, and are in the Atlantic; perhaps may enter the Streights.

I am glad to find you begin to relish Linn: there is nothing to be done in the wide boundless field of natural history without system. Now you are master of the ordines, you must attend to the genera, and make yourself well acquainted with the terms. Study well the introductions to the classes, and see how the terms are explained.

Look still for the *Myrmeleon* (lion pismier) 'Syst. Nat.' p. 913.* It has jaws like a wasp; 4 pretty long *palpi* (feelers), no *stemmata;* pimples like crowns on its head; *antennæ clavatæ!!* Andalusia, I should think, must produce it.

Your embassy to Morocco, when well drawn up, will make a good chapter in your History. Did you make no remarks on the country? You are to remember that you will want an abundance of matter to fill up 200 or 300 pages: and no publication will make a respectable

* This reference is to the twelfth edition of the 'Systema Naturæ,' of which the two parts of the first volume were published in 1766 and 1767.—A. N.

appearance unless you can swell it to somewhat of such a bulk.

What sorts of Land-tortoises do you find: when do they come forth and when do they hide?

Have you no stone-curlews (*Charadrius œdicnemus*)? they certainly leave us for some of the dead months of winter.

You will, I hope, settle that curious article concerning your winter-martin.* In your letter of November last you seemed to be puzzled, and say "that the winter-martins begin to appear in a different dress: they are blacker on the back, and whiter under the belly than last winter," and "that you suspect they are the real summer-martins now undergoing a change of colour, and possibly intending to winter here in a browner habit." And yet in your letter of April 14th you only say in general, "that you saw (March 23) swallows, martins, and your brown winter-martins all flying together." This most curious article of all your intelligence will not, I hope, remain dubious and unsettled.

Sure you must mistake when you say in your *Journal*, April 15th, 1769, "that the vines, though their shoots are but 6 or 8 inches long, have a good many grapes set." Do you not mistake the buds of bloom for fruit? Vines are late blowers in most climates: they shew the rudiments of bloom with us in April, but do not blow 'til about July 1, 'til the shoots are two or three feet long. When in bloom they smell sweetly.

Are not some of your foxes jackalls (*Lupus aureus*)? That animal wants to be better described.

Don't be too hasty in pronouncing any species a non-descript. Scopoli is very ingenious: he is publishing on birds.

* This was the *Hirundo rupestris* described by Scopoli ('Ann. I. Historico-Naturalis,' p. 167) in the preceding year, as Gilbert White subsequently found. It is the *Cotile rupestris* of modern ornithologists, the rock or crag martin. —A. N.

Mr. Pennant has heard of your pursuits, and desires to promote them. As to fishes, he says you must get Brünnich's history of those of Marseilles; and Gouan on fish: the last lives at Monpellier. Can't you contrive to correspond with him? He has written to Mr. Pennant. He expects the birds and fishes of Leghorn and Naples soon, and is ready to communicate them.

Mineralogy must not be neglected.

In order to assist your enquiries Mr. Pennant sends you a list of such animals as are known to belong to the southern parts of Europe. Your wine proves very sound and good.

In June, 1770, the Naturalist visited his relatives in London, noting in his *Journal* the occurrence of hinds on Bagshot Heath; and later in the summer received his friends, Mr. and Mrs. John Mulso, from Witney, at Selborne. Mulso proposed to proceed to Alton, where, it may be noted, he asked that "some able guide" should meet him at the Swan Inn, in order to conduct him through the devious lanes to Selborne.

Writing on September 16th, after thanking his old friend for his kindness and hospitality, Mulso continues—

"My daughter has learned such a curiosity from you about Zoology and its *Genera* that nothing but providing some 'Systema Naturæ' hereafter for her will be able to allay. If any errata should come of it you are concerned to look to them, for it is all your own doing: as to my wife I don't believe she cares a farthing about the Difference between a Penguin and the *Coleoptera*. . . . I find my garden here very disgusting to me after the extreme neatness and beautiful grace of yours."

The following letter to Pennant was naturally not included in those published, since it did not relate to Selborne.

Selborne, July 12, 1770.

Dear Sir,—A journey to London, and an other since, from whence I am but just returned, have prevented my paying that attention to your last letter that I could have wished.

If you knew how little I had to communicate to you with respect to specimens from Gibraltar * 'til I went last to town, you would not think I had neglected you: for 'til that time I had only received two *muscicapæ* and three insects. One of the birds proves, I find, to be Edwards's grey redstart; † the other, which has a white forehead, a tawny occiput and scapulars, black wings, a white rump, and black and white tail, black throat and cheeks, a tawny breast, and whitish belly—I cannot at present ascertain. ‡

The three insects were a *Panorpa coa;* rare, and peculiar in its hind wings! a large fine *Vespa*, the *crabroni congener in Italiâ capta Raij.*, *vide* 'Hist. Insect.', p. 250; and a large Ichneumon.

When I came to town I found a box containing several birds, the most curious of which are—

Merops apiaster,	Stays all the summer.
Loxia coccothraust[*es*],	Autumn and winter.
Motacilla stapazina,	Comes in autumn.
Puffin Razor-bill }	Stay all the winter.
Lanius excubitor,	Common in Spain.

* No doubt the collection sent by John White and mentioned in Gilbert's letter to him of May 26th, 1770.

† Commonly known now as the black redstart, *Ruticilla titys.*—A. N.

‡ Doubtless a species of wheatear, and the mention of the black throat shows it to have been the *Saxicola stapazina* of most modern authors, and the *Motacilla stapazina* of Linnæus was most likely the nearly allied species now known as *S. albicollis* or *aurita.*—A. N.

Charadrius calidris,	Winter and summer.
Hirundo hyberna,	Seen only in winter.
Scolopax glottis,	Common in winter.
Tetrao coturnici similis, pedibus tridactylis	Smaller than the quail, and called trail, or terraile.
Edwards's grey redstart,	Frequents the same solitary parts as the redstart.

The *Hirundo hyberna* (for so I have named it) will prove, I trust, a curiosity;* for I cannot find it among Brisson's 17 species; nor among the 12 species of the 'Syst. Nat.' It has the aspect of an *Hirundo riparia,* but seems (for I have had no opportunity of comparing it yet with our bank-martin) to be much larger, and to have a redder cast on the throat, breast and belly. Every feather of the tail, except the two midmost and the two outmost, has a remarkable white spot about midway.

If the quail should prove to be a tridactyl species, and not a variety, it will be curious.† My Bro[r] speaks of them as common; and mentions the name by which sportsmen distinguish them. For my part I think my specimen is in colour much like a common hen-quail. Brisson mentions quails in Madagascar that have no back toe; but the cocks at least of this sort have a black throat, which mine has not.

The most curious Insects in my bottles were—

Scorpio Europæus,	Cancer arctus,
A large Cicade,	Several curious Cancri; not ascertained,
Blatta Americana,	

* Subsequently found to be identical with the *Hirundo rupestris* described by Scopoli (see Letter to John White of January 25th, 1771).—A. N.

† This proves what had been before suspected (Bell, ii. p. 6, note), that John White had met with the "Gibraltar Quail," first described and figured by Latham in 1783 ('General Synopsis of Birds,' ii. p. 790 and title-page) from a specimen in the Leverian Museum, very likely obtained by John White himself. It is the *Turnix sylvatica* of modern ornithology; the *torillo* (little bull), written by him above "terraile," of Spaniards.—A. N.

Some caterpillars,
Some Scarabæi,
Scolopendra coleoptrata,
Mantis religiosa,
Spiders,
Asilus barbarus!
Onisci,
Several Labri,
*Arnoglossus, Solea lævis Raij,**
Coryphæna psittacus,

Cancer Diogenes.

Some of the fishes were—
Syngnathus acus,
Syngnathus hippocampus,
Salmo eperlanus,
Mullus barbatus,
Several Spari,
Perca marina. etc.

Sepia sepiola.

But I refer you to Mr. Barrington with respect to the fishes, who, with a person he is to procure, and my Brother in Thames-street, is to look them over more narrowly at his chambers. The specimens of fishes are in general too small; in order that they might be crouded into little room. In the autumn I expect an other box with many more specimens.

My acknowledgements are due for your list of South European animals, which I have sent, not doubting but that it will be of service; and also for ye wing of the chatterer. On your recommendation I have desired my Brother to get Brünnich, and Gouan on fishes.†

Returning you many thanks for your offers of assistance in our researches into the natural knowledge of Andalusia, which I am conscious will be very useful and necessary, I conclude

Your obliged, and
Humble Servant,
GIL. WHITE.

P.S.—Please to ascertain my second *Muscicapa.* When

* 'Synopsis Piscium,' p. 34.—A. N.
† See his letter to John White of May 26th, 1770.

I have the pleasure of meeting you I shall be glad to communicate my papers.

Since I wrote the above I have been this evening in the forest, and have procured two bank-martins, which are every way different from my *Hirundo hyberna.*

In October her nephew made his usual visit to Mrs. Snooke at Ringmer, where he noted crossbills among her Scotch firs—a fact he recorded when addressing Daines Barrington from Ringmer.*

Letter XXX., of August 1st, 1770, to Pennant, originally commenced—

"Your obliging letter of July 24th arrived last night, and I sit down this morning to answer it. I shall send you my little cargo of curiosities with a great deal of satisfaction. The birds are here at my house, but I will send them up to town to my brother in Thames Street, who has got the fishes; and will desire him to send them all together down to Chester. If you should think proper to order your artist to take any of my animals, I should be glad to see the drawings.

"When you have ascertained the fishes you will be pleased to give me an exact account of them. The birds will be labelled numerically 1, 2, 3, etc., so that you will be able to speak of them with precision."

Letter XXXI., September 14th, as originally written, concluded—

"I return you thanks for your proof-sheet respecting the elks, and am pleased to see that my description of the moose corresponds so well with yours. Last night as I rode home thro' Alton, I found at the post-house contained in three franks *Mar: Th: Brunnichii Ichthyologia Massiliensis.* My

* *Vide* 'The Natural History of Selborne,' Letter VII. to Barrington.

best acknowledgments are due for so curious and rare a present."

Letter XXXII. to Pennant, written on October 29th, 1770, contained, at its commencement, many sentences which were omitted in publication as not germane to the work. After a discussion of some of the birds forwarded by John White from Gibraltar, and now in Pennant's possession, the letter proceeds—

"You tender me Kramer* in so obliging a manner, and give so tempting a description of his Fauna, that I don't know how to waive so pleasing an offer; and yet I should be sorry to give you any trouble on that account.

"I will desire my Brother to take the height of the rock of Gibraltar: was it not stupendous there could not be such a resort and rendezvous of so many sorts of wild and shy birds amidst such a concourse of people. In an E. wind or levant the top is usually capped with a fog. On Saturday night last I was gratified with your pleasing letter of Octr 21. I mention this circumstance to show you that I lose no time in returning your fine drawings, as you desired they might not be detained. Your Artist has done my birds a great deal of credit, as well as himself; and I hope they will get back safe without any injury. The *Junco* is finely expressed and the Quail is, I think, as lovely a drawing as ever I saw. If I might object at all to any part of the performance it should be the right wing of the *Hirundo*, which perhaps is rather stiff, and to the middle of the tail, which seems too round. For the tail, though not forked, is somewhat emarginated, as Scopoli observes. The oval spots of the tail, which are characteristic of this species, are well hit off. The secondary wing feathers are, you must observe, deeply knotched.

* G. H. Kramer, 'Elenchus vegetabilium et animalium per Austriam Inferiorem observatorum.' Vienna: 1756.—A. N.

"I rejoice in your acquisition of N. American animals, and am pleased to find that you persist in additions to your 'Brit[ish] Zoology' illustrated. Such hints as occur on any of those subjects shall be much at your service.

"It gives me real pleasure to hear that the report concerning Mr. Banks is groundless. If there should be a rupture with Spain my Brother will be much circumscribed in his excursions as he has been already this summer by the death of his horse."

*From Mr. Sheffield.**

Worcester Coll., Decr. 10, 1770.

Dear Sir,—I am ashamed to have been so long in arrear for your last Favour. . . . Had I known of the arrival of your curious cargo from Gibraltar before I left the South of Oxfordshire, you had most certainly seen me again at Selborne: I could not possibly have resisted so strong a temptation. At present I must content myself with the hopes of gratifying my curiosity at a more favourable season of the year; being advised to use every Precaution, and not run the least risk of taking cold this Winter.

I had never heard of Scopoli's 'Anni Historico Naturales' but from the Description there seems to be no doubt but his *rupestris* is your Brother's *Hirundo hyberna Andal*: I lay little stress on those passages you have marked as exceptionable, having observed in Linnæus and other Systematists that the specific difference is almost universally constituted by one, or two particular marks; the rest of the Description running in general Terms—something like the expression *more* or *less* in our own Language, when referred to estimates. But I think Scopoli's Diagnosis sufficient to distinguish this species of Hirundo, & particularly the *Rectrice maculâ ovali albâ in latere interno,* without any other Description. And here I cannot help observing that Honest John Ray

* Fellow, and afterwards Provost, of Worcester College, Oxford.

is the only Describer I ever met with that conveys some precise idea in every word or Term he makes use of. Though your Brother is forestalled in the discovery, yet his ascertaining the winter residence of this new species of swallow will probably be attended with considerable advantages to natural History; as it must in great measure, if not entirely, overturn that, in my opinion, very absurd hypothesis of swallows living all Winter in a dormant state under water. I call this hypothesis absurd, as it asserts a body, bulk for bulk, specifically lighter than water, sinks to the bottom of a pool or river, contrary to a known and allowed principle in Hydrostatics—and farther—why may we not reason, by analogy, from your Brother's discovery, that those Species of *Hirundines* which frequent and affect warmer summer situations, than the summits of the Alps, require a greater degree of warmth in winter, and of course retire at their usual times of migration into more southern Latitudes? I am inclined to believe, was the experiment fairly made, that most of the known species of this *Genus* would be discovered along the shores of Africa; nor do I in the least doubt the veracity of Adanson,* when he asserts that He observed vast multitudes of European swallows in the neighbourhood of Senegal and Goree.

After I left your hospitable roof, Kew and Hammersmith were the scenes of my amusement, and highly entertained I was at each of these places, at the former with Plants, at the latter with Insects. I made a considerable addition to my collection in each of these Branches. Lee was greatly pleased with the *Panorpa coa* and the *Vespa crabroni congener*, which He had not seen before. I told Him I flattered myself I had interest with you to procure him a specimen of each, and perhaps of some others which were new to him.

* Michel Adanson, author of 'Histoire Naturelle de Sénégal avec la relation abrégée d'un voyage fait en ce pays 1749–53,' published in Paris in 1757. An English translation appeared in London in 1759.—A. N.

He would return an hundredfold, he said—indeed he has done that already to me in several instances. I must entreat you therefore to lay by a Specimen or two for Lee if you can spare them. I left the *Hirundo rupestris* with Nancy Lee to figure, received it a day or two ago drawn inimitably. I wish you much to see it. I met with a cruel disappointment in not finding Drury* at home, whose collection of Insects I went to Town almost on Purpose to see. I called at your Brother's in Fleet-street, but my friend Benjamin and he were gone to Ringmer. I met with Hudson who was civil and gave me a curious plant or two and some Insects. I have scarce room to assure you that I am with sincere regard

Your much obliged and obed[t] Hble Serv[t],

W. SHEFFIELD.

Writing on December 8th, 1770, Mulso says:—

"Have you sent my Fly to Jack [John White], and has he acknowledged the receipt? and was he pleased to see a little effort of an old friend to amuse him? Why do you not tell me of all these things, and of what he says of the Spanish camps, and the apprehensions of the garrison about it, or whether these things are the monsters of the stockjobbers? I desire you to get some spectacles; I own, as a Batchelor, that it may have an awkward look before Ladies, but I should get a great deal more out of you in half a sheet of paper, for you now write a hand so preposterously large, that one of my pages contains more than three of yours; and as you now write alone by the fireside in the evening before you go over to Mr. Etty's, you may unpannell your hose, taking care to rub the sides a little, and no one be the wiser for it but myself!"

* The well-known entomologist. Author of 'Illustrations of Natural History, wherein are exhibited upwards of 240 figures of exotic Insects, etc.,' published in London in three volumes, 1770–82.—A. N.

That Gilbert White was now really at work on his proposed publication appears from the following remarks of the same correspondent a little later in the month (December 27th, 1770):—

"As to 'Charles the 5th,' I finished him in three weeks, and you have had three months, a solitary House and a Fire to yourself: so that unless you purposely interrupt yourself in order to prolong your Pleasure, it must be finished in all this time. But you have an inexhaustible Fund in your 'Systema'! true: but as that will never be over as long as you live, I will not admit it as an excuse for not writing to me: *tædet harum quotidianarum formarum*. . . . Jacky* talks much of being your neighbour at Mr. Willis's [school] at Alton, and I bespeak him the friendship of your nephews, especially Gibraltar Jack."

The last was the only child of John White, who had, it will be remembered, been brought home in the preceding year by his mother.

The following letter is addressed to the Naturalist's sister's (Mrs. Barker's) only son, then a boy in his fourteenth year. He grew up to have many tastes in common with his uncle, with whom he constantly corresponded.

To Samuel Barker.

Selborne, Jan. 1, 1771.

Dear Sam,—I was much pleased to see so intelligent a letter from so young a writer, and shall be very glad to have you continue your correspondence.

The lines from the 'Odyssey' are very *apropos* and will make a very suitable motto for the climate of Andalusia.

* His eldest son.

My brother makes a very rapid progress in natural knowledge, and, considering he has no person to confer with or to advise him in his new study, does wonders. He sent me in October a fresh cargo of birds and insects which ought to have been here long ago; but as they came in a Levant ship, they are performing quarantine at Stangate Creek and will, I fear, be tumbled about and damaged.

When I opened your letter all the *Parnassia*-seed fell out, and I took it to be dust and dirt from the pocket of the person who brought it; but luckily it fell in my lap, so that I saved it all. I shall sow it soon in the sandy bogs, and see if I can succeed better.

The last winter migration that we have in these parts is the appearance of the *Œnas* sive *Vinago* Raii, the wild wood pigeon or stock dove, which comes in great flocks about the end of November, and does not breed in these parts, perhaps not in the kingdom.* The pigeon that breeds in our woods and hedge-rows, and cooes all the summer is the *Palumbus* or ring-dove, the *palumbes* mentioned by Virgil in his eclogues:

"Nec tamen interea raucae tua cura palumbes."

Your Uncle Harry was with me towards the end of November. As we were walking in the evening we saw just after sunset a star of a moderate magnitude, just above the sun, which we concluded must be Mercury. My Brother was much pleased to see what he thought to be that planet, as it was new to him; and I had never seen it before but once, and that was at Lyndon in 1760. You may let me know if Mercury was visible at that time.

With the compliments and good wishes of the season I conclude,

Yr. affectionate Uncle,

GIL. WHITE.

* This seems to be a mistake, as for many years the stock-dove has been known to breed abundantly in many parts of the country, and one can hardly suppose it did not breed at and about Selborne in 1771.—A. N.

At this time a letter was addressed to Pennant, which for reasons before assigned was not included amongst those published by its writer.

To T. Pennant.

Selborne, Jan. 12th, 1771.

Dear Sir,—This day my box with the whole of my curiosities, sets out by the waggon on its way to London, from whence it will be forwarded by my Brother to Chester.

You will be so kind as to examine the contents, and to order your artist to draw such as are worthy of your notice; and to favour me with your opinion concerning the most rare, and particularly the fishes, which need not be returned.

The reason that my Brother sent only the head and the feet of the vulture was because he never had any other part. The bird was found dead and floating in the sea; an accident it seems not very uncommon: some fishermen picked it up, and flayed it, eat the carcase, and threw away the skin, and gave him the head and feet. But as the Governor has got a live bird of this sort, my Brother will take care to describe that minutely.

Please to be particular about the partridges. My last cargo of birds returned very safe from your house.

I thank you for the Portugal *apiaster*, which differs somewhat from the Andalusian.

It is no small discovery, I think, to find that our small short-winged summer birds of passage are to be seen spring and autumn on the very skirts of Europe: it is a very strong presumptive proof of their migrations.

Your proof-sheet* meets with my approbation. I always was of opinion that the stile should be in some measure adapted to the length of the composition, or the subject in all cases; and therefore long flowing sentences can't be suitable to short descriptions in a work that professes to be a synopsis.

* No doubt of the 'Synopsis of Quadrupeds,' published in 1771.—A. N.

If you should think it proper to have the *Hirundo melba* taken, would it not be right to have it drawn on its back; because the colour of the belly is, size excepted, the chief thing that distinguishes it from the *Hir. apus.* It is a swift to all intents and purposes.

You pay us a great compliment when you say that our account of Gibraltar will in a manner comprehend the animals of South Europe. It is a work that I could wish see reduced to some degree of correctness; and therefore am mnch gratified when you tender us your best assistance, which I am perswaded would be its best support. I have been in a pother lately about writing to that place, fearing lest this misunderstanding between the two nations may interrupt the correspondence by the post, and suspecting my last letter never reached my Brother. Mr. Barrington seems to think that the intercourse is still subsisting.

I have looked over Mr. Forster's catalogue of British insects,* and have somewhat to advance on that subject; but time will not permit me at present, as I am pretty much hurryed. We have had vast rains for these ten weeks past, and some great storms, especially one on the 20th December in the morning; now severe frost.

Hoping your troublesome cold has left you, and desiring you to accept the good wishes of the season,

I conclude with great esteem,

Your obliged and humble servant,

GIL. WHITE.

To the Rev. John White.†

[Headed "Bro. John."]

(2)

Selborne, Jan. 25, 1771.

Dear Brother,—I received your kind letter of October 19th, and wrote you an answer on November 6th. I should

* Published at Warrington, under the auspices of Pennant, in 1770.—A. N.

† At Gibraltar.

have been very glad to have seen Mr. Twisse: he just came to London, called on Bro. Ben, and set out for Gibraltar again. N° five is Ray's *Junco*, and the *Turdus arundinaceus* of Linn. The *Merula passer solitarius* of Ray is said to be a fine songster, and is supposed to be the bird mentioned in Psalm cii. 7. Your winter swallow is undoubtedly the *Hirundo rupestris* of Scopoli; you however will have the credit of discovering its winter quarters. Brisson mentions a tridactyl quail from Madagascar. He calls it "Perdix infernè cinerea, supernè e cinereo, rufo, & nigro variegata, gutture & collo inferiore nigris; coturnici nostrati *paululum crassitie cedens.*" His 'Ornithology' is extravagantly dear: 7 or 8 guineas. Geoffroy will set you right by means of his cuts in many genera of insects.

The motto from the 'Odyssey,' book iv. 566, is a description of the Elysian fields, and will suit the climate of Andalusia well—

> "Οὐ νιφετὸς, οὔτ' ἂρ χειμὼν πολὺς, οὔτε ποτ' ὄμβρος·
> 'Αλλ' αἰεὶ Ζεφύροιο λιγυ πνείοντας ἀήτας
> 'Ωκεανὸς ἀνίησιν αναψύχειν ἀνθρώπους."

> "Stern winter smiles on that auspicious clime;
> The fields are florid with unfading prime:
> From the bleak pole no winds inclement blow,
> Mould the round hail, or flake the fleecy snow;
> But from the breezy deep the blest inhale
> The fragrant murmurs of the western gale."

The prose motto is perfectly suitable to your present situation, and prophetic of your undertaking.

"Certe si aliquis Naturæ consultus in maxime australi Hispaniâ in aves observaret, quando accedant aut recedant austrum et septentrionem versus, notatis scilicet diebus mensis et speciebus; res haec adeo obscura brevi maxime illustraretur." 'Amœnitates Academicæ' Lin., vol. iv.

Now Mr. Twisse is returned, be sure get his conjectures on the currents of the Streights: you will want dissertations for your work. Your embassy to the Emperor of Morocco with a description of his person, manners, troops, etc. will make a very good chapter. Have you not in Spain some crown-flocks of sheep which migrate with the seasons from N. to S. Get some anecdotes of them. Mr. Pennant makes his artist take all your most curious birds; and promises the drawings shall be forthcoming if wanted to engrave from. Describe the *Vultur percnopterus* most minutely, and learn if you have an opportunity the difference of the sexes. Get the skin of the *Lupus aureus* from Barbary, and describe it well. Scopoli's new *Hirundo alpina* is nothing, I think, but the *Hirundo melba*, which is indeed a noble swift: get all the anecdotes you can about them. Write to Scopoli, he is very clever: but ask him as gravely as you can how he is sure that the woodcock, when pursued, carries off her young in her bill.* I have just sent your cargo, which I received in August, to Mr. Pennant: but as to your collection shipped in October, I have never seen it yet; for brother Thomas writes word that it has been performing quarantine in Stangate-creek. Just as I had penned the last sentence a letter arrived from brother Thomas informing me that the box was got safe to his house, which is good news: for I was in pain for the curiosities and Jack's shirts. When the *Mantis* casts his skins he is in his pupa state, and advancing by casting aside those exuviae to perfection. Scopoli's *Icones* will probably disappoint you: Linnæus's engravings of insects are miserable: Geoffroy's are the best I have seen. The bird you call a *Parus* (if it be not the common black-cap) is a nondescript: if it should prove new, call it *Motacilla atricapilloides:* Mr. Pennant thinks it a new bird. Your purple-winged *Vespa* is no doubt the *Crabroni congener* Raij; and if you can find that it has "thorax ad latera postice utrinque dente notatus," I shall

* Scopoli's words are "pullos rostro portat," which should be rendered "by means of her bill," and relate what is since proved to be a fact.—A. N.

acknowledge it to be the *Sphex bidens* Linn. I am sending all your insects to nephew Ben. White in town, and shall get Mr. Lee, the botanist of Hammersmith, to inspect and ascertain them, because he is the best Entomologist that I know. The reason that Linn. mentions so many Insects from Barbary is, because Mr. Brander, the Swedish Consul at Algiers, sent him vast collections. In the little box which you sent me with a sliding lid are two species of *Myrmeleones*. Geoffroy seems to have a good cut of one. You will now be able to measure the rain of your climate: the mean quantity *per ann.* in Rutland is 20¾ inches. Learn as much as possible the manners of animals, they are worth a ream of descriptions. You must produce some ingenious dissertations to entertain the unsystematic reader. What do the *Panorpæ coæ* do with their long remiform wings? Frequent your markets, and see what birds are offered to sale. Get some account of the prickly heat, or fever, and the exact height of your mountain. It seems to me no doubt but that your *Motacilla* N° 5 is the *Junco* of Ray: but he does not seem to be so exact as usual, and talks of a stiff tail, and omits mentioning the white and black bars at the end of its tail. It will be worth our while to find out Mr. Moore, the botanist, or his representatives; and to endeavour to procure his flora of your district. Ray does not take notice that the thighs of the *Merops* are naked.

I had written thus far when your curious box of birds shipped in Oct[r] and Jack's shirts and sweetmeats arrived: the insects were left in town for the reason above-mentioned. Your kind letter of Dec[r] 9 came the same day. Geoffroy, no doubt, is too verbose, so are all his country men. Mr. Pennant makes sad complaint of Gouan's book of fishes, and of the obscurity of the *Labrus* and *Sparus* genera. D[r] Shaw's Natural part of his travels * is said to be good. You will

* 'Travels in Barbary,' etc.—A. N.

do well to have two columns of thermometer observations, especially as 1769 and 1770 were both on the extremes. As matter flows in upon me I begin to think of composing a Natural History of Selborne in the form of a journal for 1769: we shall then be able to compare the climates. You mention the great eagle owl, and send me, I think, a wing and claw of that majestic bird; and yet you call it *Strix otus;* sure you mean *Strix bubo;* the *otus* is our common horned owl. I see none of your plants; perhaps they are lost: the sweet-smelling clammy shrub must be, I suppose, a cistus; has it not a single, rose-like, fugacious flower? You have classed all your last fine cargo of birds so justly, that there is no room for objection. Where you doubt, I doubt: though I think there is little room to doubt about the *Alauda cristata:* but the pair of birds (if they are a pair) which, I suppose, you call *Alauda non cristata,* seem rather some species of the genus of *Motacilla.* Get the *Pratincola* when you can. At present I am a stranger to your *Œnanthe.* The *Oriolus galbula* must be a fine bird when in perfection. Your barometer fluctuates much more than I could have expected in so warm a climate and low a latitude: in the tropics it hardly varies at all. Your last quail seems to be a male, the former a female. You will pardon the didactic air of my letters, which in our present way of correspondence is perhaps unavoidable. The wing of the *Strix bubo* is "*remigibus primoribus serratis*": had Linn. remarked that, he would not have made that a specific difference to his *Strix aluco.* See 'Fauna Suec.,' p. 25.

I am, &c., &c.

CHAPTER X.

Early in the year 1771 Mrs. Woods died. The loss of this sister seems, from a letter of Mulso's, to have been greatly felt by Gilbert White.

On March 30th he addressed to Pennant what became Letter XXXIV. The original letter commenced—

Your agreeable letter of Feb. 15th arrived while I was from home, and since my return somewhat still has prevented my sitting down to pay it that regard which it deserved.

You may probably have heard by means of Mr. Barrington, who saw the contents while they lay in town, that I have received another small cargo of birds from Gibraltar, with a curious collection of insects. The birds were as follow:—

Merops apiaster, 3 specimens,
Loxia curvirostra,
Scolopax œgocephala, M. & F.,
„ *phœopus*,
Oriolus galbula, 2 spec.,
Alauda cristata,
Alauda „ ? 2 spec.,
Coturnix tridact, Mas,
Mot. œnanthe, M. & F.
Rallus aquat.
Motacilla flava.
Œnanthe ?
Charadrius hiaticula.
Hematopus ostralegus.
Leg & wing of *strix bubo.*
Leg of *ardea nycticorax.*
Turdus arundinaceus.

Where a wing or a leg or an head only are sent, you are to suppose that the whole specimen was too stale and too far gone to be preserved before it reached my Brother's hands. The *alauda* unknown answers well in many respects to the *Spipoletta Florentinis* of Ray. But as that most accurate writer says that the *rostrum* of the *spipoletta* is *nigerrimum*, and *pedes etiam nigri*, I must by no means pretend to say that my birds are the above-mentioned when their bills and legs are brown; and especially since all ornithologists agree that the naked parts of birds are the least apt to vary in colour. As to the *œnanthe* I don't know at all what to make of it; it appears to me more like a variegated accidental specimen than a new species: but I shall hear what you have to say. The outer edge of the first quill feather of the wing of the *strix bubo* is serrated, a circumstance which Linn. seems not to be aware of; for if he had he would never have made it specific to his *strix aluco*: since what is common to more than one species cannot be specific. But such slips are pardonable, nay unavoidable, *opere in longo.*

As the *orioli galb.* are birds of last year their colour is by no means come to its full splendor. My Brother has much to say in defence of his supposition that his Spanish and Barbary partridges are different species. In one of his last letters his words are, "I am perfectly clear about the difference of the Span. and Bar. partridge. I have examined multitudes of each, and never found the least exception to my remark.—That the Bar. sort has always the chestnut collar, cheeks, &c., spotted with white; the Span. sort always has those parts black, and the collar of a different form. The distinction is invariable; and I wonder no one remarked it. The Span. is rather the larger bird. Indeed on a careful comparison the whole disposition even of those colours which correspond in each bird differs."

Shaw's 'Travels' are to be met with in Gibraltar, and my

Brother had discovered himself that the tridactyl quail was known to the Dr in Barbary; however we are equally obliged to you for your hint. Gannets are never seen about Gibraltar 'til Nov.; they retire again about March. My Brother shall try to procure the bird for you from the Barbary Coast.

I shall make a point of meeting you in town. It is time now to have a little conversation face to face after we have corresponded so freely for several years.

The letter concluded—

If you have a desire to see my last birds, please to intimate as much; but as you intend soon to be in town, might they not as well meet you there and save a large carriage? but this shall be as you please.

I had written thus far when your letter of the 19 of Mar. arrived. Many thanks are due for your trouble in ascertaining so many of my Brother's fishes, and for the honour you have done his birds in ordering so many of them to be taken. I shall transcribe your list and send it off to Gibraltar next week. My brother will be pleased to see how you have named his specimens.

When you write to Gibraltar, crowd your letter with hints: mine run of late in a very didactic style. You have, I find, made some alteration in your time of coming. May I presume to ask how long you stay in town? Hoping to have the honour of seeing you soon,

I remain with great esteem

Your obliged and

humble Servant,

GIL. WHITE.

From the following letter to Pennant, written at this time, the Naturalist extracted Letter XXXV.

little more than a paragraph (here omitted) upon the tails of peacocks:—

Selborne, July 19, 1771.

Dear Sir,—My unusual silence has not been owing to any disrespect, but to the roving unsettled life which I have lived for this month past.

I wish you had happened to have paid a little more attention to the pair of larks which came over in my last collection, because they seemed to me to be quite a different species from any sent before: and I should not have hesitated to have called them *Spipoletta Florentinis Raii*, had they had black feet and black bills. The variegated *Œnanthe* also deserved your regard. But I will endeavour to send both sorts again when I have an opportunity, that you may survey them at your leisure. My thanks are due for your setting us right where some birds were misnamed.

It is a great satisfaction to me to find that you and my brother at Gibraltar are embarked in a correspondence. You are capable of giving each other mutual entertainment: and my brother (as by much the youngest Naturalist) will derive from you much information and many useful hints and queries. What from his natural propensity and application, from the assistance of ingenious friends, and from the copious field of the South of Spain, which he has all to himself, I doubt not but that in time he will be able to produce somewhat worthy the attention of men who love these studies.

As to any publication in this way of my own, I look upon it with great diffidence, finding that I ought to have begun it twenty years ago. But if I was to attempt anything it should be somewhat of a Nat[ural] history of my native parish, an *annus historico-naturalis* comprizing a journal for one whole year, and illustrated with large notes and observations. Such a beginning might induce more able naturalists to write the history of various districts, and might in time occasion the production of a work so much to be wished for, a full and compleat nat[ural] history of these kingdoms.

Your engraver at Chester acquits himself like an able artist: and I should be glad to know what his price is for a plate containing two or three animals. You have, I see, furnished the 'Gentleman's Magazine' for last month with a plate and some descriptions. The conductor of that publication will, no doubt, rejoice in such a correspondent. . . .

I have just read with satisfaction and improvement Kalm's journey thro' N. America; but as he is continually referring to another work he cuts us very short often times both in botany and zoology.

Yesterday I had a letter from town, which mentions the safe return of Mr. Banks,* and adds that he looked as well as ever he did in his life. So agreeable an event calls for my warmest congratulations. For if we rejoice at the arrival of a friend who has been absent but a few months perhaps in a neighbouring kingdom; how shall we express ourselves when we see one restored as it were from the other world, having undergone the astonishing hazards and dangers that must attend the circumnavigation of the world itself!!!

I have great reason to regret my disappointment of not meeting you in town: but as we live by hope I trust that I shall be more fortunate another time.

With great esteem I remain

Your obliged, and

Most humble Servant,

GIL. WHITE.

On September 25th, 1771, the Selborne Naturalist returned to the subject of the birds of Gibraltar—

* From Cook's first voyage.

a part of a letter to Pennant (Letter XXXVII.) omitted on publication.

By the next return of the waggon I shall send up a small but rare collection of birds, which I beg that you and Mr. Banks would please to examine, that I may hear what two such curious Naturalists have to say about some of of them. They are as follow:—

Merula passer solitarius.	M. & F.
Merula nigerrima uropygio rectricibusque niveis.	Is this not a nondescript?
Fringilla petronia.	
Sturnus collaris Scopoli.	An elegant bird!
Emberiza cirlus.	
Pratincola Krameri.	Well engraved in Kramer.
Anas clypeata pectore rubro.	Differs from our shoveler.
Species of lark.	What?
Motacilla boarula Scopoli.	
Species of perch.	What?

You are welcome to take these birds into the country, as you say London affords you no leisure for examination in such matters: and if you shall think them worthy of being drawn, you will lay us under great obligations by communicating those drawings at a proper season. My Brother makes no contemptible progress in Nat. history, and will be able, I trust, by the assistance of good friends (to whom he will be ready to make all due acknowledgements) to produce in due time somewhat not unworthy the attention of the candid Naturalist. Please to return the birds to my Brother in Thames street as usual. It will not be in my power to meet you in London at present, because I have a call that obliges me to go an other way.

Pray present my humble respects to Mr. Banks, and tell him I heartily congratulate him on his safe return from his astonishing voyage! The world expects great information from his discoveries during his circumnavigation. My respects also wait on Mr. Barrington, and thanks for his letter from Beaumaris.

After returning you my acknowledgments for your present of the curious old *Rondeletius de piscibus*,* I remain

With great esteem
Your much obliged, and
Humble Servant,
GIL. WHITE.

After entertaining some friends in the summer at Selborne, the usual autumn visit to Ringmer was made. The *Naturalist's Journal* of November 1st records that—

"Mrs. Snooke's tortoise began to dig in order to hide himself for the winter."

That swallows, as well as tortoises, might perhaps hide themselves during the winter season remained, as successive entries in his *Naturalist's Journal* distinctly show, a favourite speculation of Gilbert White's till the end of his life. At this time he notes—

"Nov. 4th. Saw three swallows flying briskly at Newhaven at the mouth of the Lewes river."

On his journey home by way of Chilgrove—

"Nov. 13. Saw 16 forktailed kites at once on the downs."

In November, 1771, John Mulso, who had recently become Canon of Winchester, received the rectory of Meonstoke from his uncle, now Bishop of Winchester, giving up the living of Witney; so that he

* There are several editions of Rondelet's work on Fishes, the first published at Lyons in 1554.—A.N.

Walker & Cockerell, ph. sc.

Reb: Snooke

now had a home only about seventeen miles south-west of Selborne, on the Alton and Gosport road. No preferment, other than those of his College, was offered to the Curate of Faringdon; and his friend, regretting this, wrote on February 27th, 1772 (after requesting good offices at an Oriel Fellowship election)—

"It is a sincere grief to me, that my perpetual desire of serving you, and my efforts towards it, have had so little success as to give me no claim to ask favours of this nature of you. . . . I hear nothing of West Meon, but that I am *not* to have it. I do not *ask*, but I might be *prevailed on* to take it. You are morally sure that *I* have not the disposal of any of the Bishop's preferments."

On January 13th, 1772, the following letter was addressed to Pennant from Selborne, but not included in those published:—

Dear Sir,—I sent you by the return of the Alton waggon last week such birds of my last cargoes as you had not seen before: some of which, I think, will not displease you; and of others I shall beg your friendly information, not being able to ascertain them, especially the *larks*, and the *motacillæ*. You will, I hope, also give me your opinion of the last cargo; and especially of the *white-rumped* bird and the duck; the former of which is, I trust, a *turdus*, and a rare bird, and perhaps a nondescript; and as to the latter I should be pleased to know whether it be the red-breasted shoveler of the Brit. Zool. or not. My present cargo is as follows:—

Phœnicopterus ruber Mas.
Larus fidipes alter Willugbœi: N. VI. ?
Lanius collurio, pullus ?

Lanius collurio, woodchat from Tetuan.
Sturnus niger from Tetuan: *an nova species?*
Oriolus galbula.
Oriolus „ young or variety.
Alauda cristata.
Alauda No. 1. 2 specimens; bills long and slender; breasts tinged with yellow.
Alauda No. 2. 2 spec.; bills short and taper; back claw small—and short; tails short and dusky; outside feathers tinged with yellow.
Alauda? bill slender; back claw short and rather incurved; breast a little spotted; tail long and dusky; out feathers white; is it the same with No. 1?
Motacilla No. 1. What? large: back and wings ash-coloured; head dusky; throat, breast, belly white; tail lost *an Motacilla dumetorum Linn. Kram. Aust.* 377:n:19?
Motacilla No. 2. What? small: head, back dusky-reddish; wings dusky; outer webs chestnut-coloured; throat white; breast tinged with red; feet and legs palish.
Motacilla No. 3. 2 spec.; minute; head, back, wings black; feathers edged with chestnut, resembling the *passer torquatus*; throat white, breast, belly, sides tawny; *caudâ unicolore*; elegant little birds?

In all these difficulties your obliging disposition will prompt you to assist me; and you will besides I hope refer so to numerical remarks as to prevent mistakes or misapprehension. It is very remarkable that of all the larks my Brother has procured he has never yet met with a British species. Has Brisson any larks unknown to Ray? You will find, I think, in the box two or three distinct and unusual species. My Brother in Thames Street has sent for the last bottles of fishes. Enclosed with the birds are some rough draughts of some fishes taken by my Brother at Gibraltar, who, though he knows nothing of the rules of drawing, yet he trusts such rude sketches will inform an Ichthyologist better than mere words.

As I have a few shells and fossils, I should take it as a favour if you would (when you return the box) add a few ores and fossils to my collection such as your mines and neighbourhood afford; a few will be sufficient.

Pray, when in this Spring do you intend to be in town? Still I hope to meet you there some time or other.

I am much pressed at present, and must stop here, but propose to write again not long hence. Pray, when does Mr. Banks sail? *

On March 19th, 1772, another letter on the same subject to Pennant was written from Selborne, but not published.

Your obliging letter of Feb. 21st came safe to this place and followed me up to town; where I also received your favour of March 1st.

While I was in London came from Gibraltar a box containing (besides several birds which you have seen before) *Ardea alba minor*; perhaps the 6 of Ray's *synopsis avium*: *Charadrius alexandrinus.* These are all the new birds.

In a bottle *Sparus mœna?* *Salmo eperlanus Calpensis*: *blennius supercilios*: *cancer arctos.*

In a phial *Squali fœtus*: *cancer arctos*: *labrus.*

These are all left with my Brother Tho. who will add them to the cargo I am sending up.

I also looked out the *pratincola*, which will be sent with the rest. There can be no doubt of its being a genus *per se.* When I came home I found by the Liverpool frigate a box containing:—

Mustella lutra. *Trigla volitans*: and
Squallus glaucus. some birds seen
„ *mustelus.* before; all dryed.
Uranoscopus scaber.

* Banks, disappointed of a second voyage round the world, in this year undertook one to Iceland.—A. N.

PHIALS.

No. 1. *Gasterost. ovatus? Zeus aper*; *Labrus*; *Perca?*
„ 2. *Esox Saurus.*
„ 3. *Cancer Squilla carinata*; *Percæ*; *Gobii.*
„ 4. *Cancer squil. carin.*; *Trigla verticillata*; *Perca.*
„ 5. *Trigla lucerna*; *Trach. Draco.*

You will also receive the outlines of the following fishes:—
Squalus centrina: *Sciæna? Borrico minor*; *Scomber pelamis*; *Sciæna*; *Corbo*; *Esox Saurus*; *Gasterosteus Saltatrix*; *Lepidopus*; *Perca vel Zeus? novus capite diaphano.*

Among the rest I send you the short-eared owl of 'British Zoology' omitted before.

My thanks are due for your thoughts on the former cargo.

Your tour through Scotland appears to me to be a very engaging work; and the town it is plain, is of the same opinion; for the book has a great run.

I regret that I was obliged to leave town before I had seen your *genera avium*.* Your *Synopsis quadrupedum* gives me satisfaction.

When I came to town I found a long letter from Linnæus to my Brother John lying in Fleet-street, occasioned by an epistle and some phials of insects sent by the latter to the former. The old arch-naturalist writes with spirit still; and is very open and communicative, acknowledging that several of the Insects were new to him. He languishes to see a *pratincola*, being conscious that it belongs not to the genus of *hirundo*.

Please to order the fishes that are ascertained to be thrown away; I mean those in spirits.

I am, sir,
Your most obedient and
Humble servant,
GIL. WHITE.

* This must have been in MS., as the first edition of the book was not published till the following year. Copies of it are exceedingly rare.—A. N.

In the spring of the year 1772 John White was presented to the vicarage of Blackburn, Lancashire, doubtless through the influence of the Governor of Gibraltar, Lieutenant-General the Hon. Edward Cornwallis, brother of the Archbishop of Canterbury, in whose gift the living was. Mulso wrote on May 26th to his friend that Blackburn, which had belonged to a former neighbour of his, was "a very good living, but overcharged with duty, and therefore eaten up with curates." He went on to the apposite remark, as it unfortunately turned out, that John White ought to have been placed "more Southerly after 16 years residence in Andalusia."

The *Naturalist's Journal* records John White's movements at this time with affectionate interest.

"June 17th; Brother John set out on horseback for Cadiz—19th, arrived at Cadiz—21st, Brother John sailed from Cadiz for England—July 27th, Brother John arrived at Gravesend in 37 days from Cadiz."

In June Mulso writes with more particulars of the Blackburn emoluments.

"I hope to hear that your brother and sister are safe arrived and well very soon. Poor Jack Gibraltar must cast a longing eye towards the sea, and you too must rejoice in seeing a brother so long removed from you. I wish you a happy meeting."

A little later Harry White also received a living, the small vicarage of Uphaven, Wilts, which he held with his Fyfield Rectory.

To Samuel Barker.

(With an extract from a letter of Mr. Sheffield's as follows—)

Decr. 21, 1772.

"My next Scene of Entertainment was in new Burlington Street at Mr. Banks's. Indeed 'twas an invitation from this gentleman that carried me to town. It would be absurd to attempt a particular description of what I saw here, it would be attempting to comprize within the Compass of a letter what can only be done in several folio volumes. His house is a perfect museum, every room contains an inestimable treasure. I passed almost a whole day here in the utmost astonishment, could scarce credit my senses, had I not been an eye-witness of this immense magazine of curiosities, I could not have thought it possible for him to have made a twentieth part of the collection. I have excited your curiosity, I wish to gratify it. But the field is so vast, and my knowledge so superficial, that I dare not attempt particulars. I will endeavour to give you a general catalogue of the furniture of three large rooms; first the Armoury; this Room contains all the warlike instruments, mechanical instruments, and utensils of every kind made use of by the Indians in the south Seas, from Terra del Fuego to the Indian Ocean. Such as bows and arrows, darts, spears of various sorts and lengths, some pointed with fish, some with human bones, polished very finely and very sharp, scull-crackers of different forms and sizes, from 1 to 9 or 10 Feet long, stone-hatchets, chisels made of human bones, canoes, paddles, &c. It may be observed here that the Indians in the south Seas were entire Strangers to the use of iron, before our countrymen, and Monsieur Bougainville, arrived amongst them; of course these instruments of all sorts are made of wood, stone, and some few of bone. They are equally strangers to the other metals: nor did our adventurers find the natives, of this part of the globe, possessed of any one species of wealth, which could tempt the polite Europeans to cut their throats and rob them. The second room contains the different habits and ornaments of the several Indian nations they discovered, together with the raw materials of which they are manufactured. All the garments of the Otaheite Indians, & the adjacent islands are made of the inner bark of the *Morus papyrifera*, and of the bread Tree, *Chitodon altile;* this cloth, if it may be so called, is very light and elegant, and has much the appearance of writing paper, but is more soft and pliant; it seems excellently adapted to these climates; indeed most of these tropical islands, if we can credit our friend's description of them, are terrestrial paradises. The New Zelanders, who live in much higher southern latitudes, are clad in a very different manner; in the winter they wear a kind of mats made of a particular species of Cyperus grass. In the summer they generally go naked, except a broad belt about their loins, made of the outer fibres of the cocoa nut, very neatly plaited; of these materials they make their fishing lines both here and in the tropical isles. When they go upon an expedition, or pay or receive visits of compliment, the chieftains appear in handsome cloaks ornamented with tufts of white Dog's hair; the materials of which these cloaks are made, are produced from a species of *Hexandria* plant very common in new Zeland; something resembling our Hemp, but of a finer harl, and much stronger, and when wrought into garments is as soft as silk; if the seeds of this plant thrive with us, as probably they will, this will be perhaps the most useful discovery they made in the whole voyage.

"But to return to our second room. Here is likewise a large collection of insects; several fine specimens of the bread & other fruits preserved in spirits; together with a compleat *hortus siccus* of all the plants collected in the course of the Voyage. The number of plants is about 3000; 110 of which are new genera and 1300 new species, which were never seen or heard of before in Europe. What raptures must they have felt to land upon countries where everything was new to them! Whole forests of nondescript trees clothed with the most beautiful flowers and foliage; and these too inhabited by several curious species of birds, equally strangers to them. I could be extravagant upon this topic; but it is time to pay our compliments to the third apartment. This room contains an almost numberless collection of animals; quadrupeds, birds, fish, amphibia, reptiles, insects, and vermes, preserved in spirits, most of them new and nondescript. Here I was lost in amazement; and cannot attempt particular description. Add to these the choicest collection of drawings in natural history that perhaps ever enriched any cabinet public or private; 987 plants drawn & coloured by Parkinson, and 1300 or 1400 more drawn, with each of them a flower, a leaf, and a portion of the stalk coloured, by the same hand, besides a number of other drawings, of animals, fish, birds, &c. And what is more extraordinary still, all the new genera and species, contained in this vast collection, are accurately described, the descriptions fairly transcribed, and fit to be put to the press. Thus I have endeavoured to give you an imperfect sketch of what I saw in new Burlington Street, and a very imperfect one it is."

Dear Sam,—As I have promised for some time to write to you without fulfilling my promise, I shall, by way of making you some amends, send you the above-written extract from my friend Mr. Sheffield's letter, instead of something of my own composing. When you and I happen to meet, we will, if you like, read Virgil's Georgics together, with Martin's translation and notes; and shall, I trust, derive no small satisfaction from that most beautiful of all human compositions. Ben White,* while with me, read them through

* Eldest son of Gilbert White's brother Benjamin.

three several times; but he was at that time almost too young to relish so masterly a work.

Give my respects to your father, and tell him I owe him a letter, which I intend to pay him soon; and inform him that hitherto our winter has been remarkably mild; within a fortnight I have cut grass for my horses; and nasturtiums abroad are still in bloom! Our mornings and evenings are full as mild now as they usually are at this season at Gibraltar: though at noon the thermometer is much raised by the sun at that place. My thermometer yesterday morning stood at 52! As I have some suspicions about the regularity of my barometer, pray send me a journal of your barometer for any month past; and let me know if the *surface* of the quicksilver in the receiver of your barometer be exactly 28 inches beneath the lowest mark on the plate.

My Brother and Sister John (who have been with me about a fortnight) are much favoured by this delicate weather, and will, I hope, be tolerably seasoned before severe frosts set in. Brother John is frequently incommoded by hoarseness, an infirmity that is very troublesome to a clergyman.

Jack White is here, and is my amanuensis. The whooping cough is crept in among nurse Butler's children; but Nanny Woods has not yet got that troublesome complaint.

You will, I hope, write soon, and let me hear how you succeed in your studies; and how much you and your sisters improve in drawing, and particularly in designing. Your sentiments on any subject will be very agreeable to me. All friends join in respects.

I am y^r affectionate friend,

GIL. WHITE.

Mr. and Mrs. John White spent the winter of 1772–3 at Selborne, during which time the former continued a correspondence with Linnæus, commenced at Gibraltar in June, 1771. On January 10th Mulso writes:—

"I suppose your brother John and his Lady are put up in cotton with some of the Andalusian rarities, for how they can stand against this severe weather after their broiling on the Rock so long I cannot imagine. I very much long to see them: should you stare very much if you saw me come tottering down the hill supported by my man, and would you allow me, as a Witney man, an additional blanket? and, as an old soaker, a double portion of your spirits? For I fancy the north side of the Hanger to be pure and cold, and to demand every *Succedaneum* to comfort and warmth. . . . I reckon you have pure, hospitable, Christmas doings in your three neighbourly families.* Be jovial and refresh your hearts, and forget not in your cup your old and faithful friend,

J. Mulso."

The following letter was written by John White during his stay with his brother, on the same sheet as one of the letters to Pennant, which became Letter XXXVIII. of 'The Natural History of Selborne.'

To T. Pennant.

Selbourn, Mar. 16, 1773.

Dear Sir,—I am sorry our affairs do not coincide a little better, so as to give us an opportunity of meeting in London. According to our present plan, my brother and I propose to be there about the middle of April, at which time it

* *i.e.* besides the occupants of "The Wakes," the Ettys at Selborne vicarage, and the Yaldens at Newton Valence vicarage.

is feared to be you will be returned into the country. I am obliged to you for your kind offer concerning the drawings, but cannot yet positively say which in particular I would wish to have copied. I must first consult some of my friends on that head, and should be glad of your opinion in the choice of them. Linnæus says the fish, which I am doubtful whether to call *perca*, or *zeus*, is actually a *new genus*! I have a good specimen which you shall see; and it will be better to draw from that than my rude outline. I shall now be glad to collect *all* my scattered remarks on the Nat. Hist. of Gibraltar, and shall beg the favour of seeing, once more, those anecdotes which I have sent you from time to time, especially those that relate to the fishes and birds. I beg your thoughts on the *lepidopus*. It certainly is what Gouan speaks of, though very different in some respects. I shall have a great variety of new insects, but I fear many more are lost by being too hastily handled and examined. On examining the biggest of my short-billed Andalusian larks, and comparing it with the British larks here, I find it to be the real and genuine skylark *Alauda arvensis*. However, I hope it's past all doubt that I have still *two new larks*. What think you of *Gasterosteus saltatrix* as I have ventured to call it? Yet I fear it cannot be positively pronounced that species. I would like to have cuts of all my *new* subjects, provided they were well executed. But I observe that all the artists, who succeed pretty well in quadrups. and fishes and insects, are still very defective in birds. Have you and M. de Buffon adjusted matters concerning the *Scops*? I have a pretty specimen, but fear the engravers will murder the delicate pencilling of that bird. I should be happy to have some days' conference with you on the more rare of our subjects, before I finish my Fauna. I am not so forward as I could wish for want of having all my specimens and materials about me. If you have any papers or memorials that you can spare which may contribute to my farther information, please to leave them at my brother's in Fleet street. I hope to be at Blackburn the beginning of May.

I am, D[r] Sir,
With much esteem,
Y[r] most obed[t] serv[t],
J. WHITE.

In your list of Animals of Southern Europe which you have got drawn, I cannot precisely distinguish which of them are mine, but shall be glad if you would put a mark on those which you had from me.

A little later the *Naturalist's Journal* records a visit to London with Mr. and Mrs. John White, John White being then very likely engaged upon the business of his proposed publication of his 'Fauna Calpensis.' Gilbert White returned alone on May 21st.

To the Rev. John White.

Selborne, June 26, 1773.

Dear Brother,—Your favour of the 17th reached me last Wednesday; and about the same time, I presume, you received my account about Jack's measles. My nephew continues perfectly well, and has not through the distemper nor since had the least cough. From the time that he came home he had somewhat of an hoarseness in his voice, which I took at first to be a cold; but upon considering the matter, it is owing no doubt to a cause incident to young men about his time of life.

It pleases me much to find that you have heard the sibilous, or shivering wren;* since now you know all the

* The wood-wren or warbler of modern English writers.—A. N.

species; and that you have heard the sedge-bird; which for variety of notes, and swift transitions from the song of one bird to that of an other is, I think, a wonderful fellow, and was it not for the hurrying manner, would be an elegant warbler. It is plain Mr. Lever knows nothing of the grasshopper-lark; if he did, he could not confound it with the sedge-bird, to which it bears not the least resemblance either in person, song, or manner of life. Did the shivering wren make its noise in the tops of tall trees? Mr. Lever is, I perceive, a very adroit *natural* Naturalist; it is therefore pity he does not allow himself the advantage of books, and call in the assistance of system.

The sedge-bird sings all night when it is awake, therefore when you throw stones or dirt into the bushes, you rouse it from its slumbers, and set it to work again.

You will be very busy, no doubt, in your repairs, and will meet, I hope, with no disappointments. I thought a fortnight ago that I was going to build a chamber full speed. I had bespoke a mason in the room of Long, who was pre-engaged; and Jack was to have been the Comptroller general of my Majesty's works; but just as I was going to lay in all materials my Mason sent me word he had got an other job, and could not do mine 'til *after harvest.*

For these three days past we have had the king at Portsmouth; and have heard continual firings, which shook my house. My s[t]foin is down; but the weather is unsettled. Mr. Lever has procured the *canne petiere** in Lancashire; Mr. Pennant mentions one shot in Cornwall. Some boys killed lately at Oakhanger-ponds some flappers or young wild-ducks; among the rest they took some young teals alive; one I saw, and turned into James Knight's ponds. 'Til now I never knew that teals bred in England. So you see new information crowds in every day. Was not the sibilous bird that you heard the real grasshopper-lark;

* The little bustard (*Otis tetrax*).—A. N.

did it haunt the tops of the tallest trees, or low bush-hedges; did it sing by night o by day? Many children continue to die of the measles, among the rest the youngest of Mrs. Hale's this morning; and the whooping cough rather gets worse than better. Poor Nanny Woods's cough is very bad; and she is very weak, and mends very slowly.

Mr. Knight of Street house is dead.

With respects to my sister, I remain
Your affectionate, and obliged brother,
GIL. WHITE.

Earlier in the year Mrs. Chapone had published her 'Letters on the Improvement of the Mind,' which were written for the instruction of her niece, Miss Mulso. Though Mrs. Chapone is best known to the present generation as an object of Thackeray's satire, her brother, John Mulso, was able to write to his friend, dating from Meonstoke, July 5th, 1773—

"My sister Chapone . . . I have not seen since the great harvest of her fame. She is much gratified by the praises that resound on all sides, and indeed I fairly think that she deserves them."

To the Rev. John White.

Selborne, Aug. 2, 1773.

Dear Brother,—I find you still, as well as when you resided on the other side of the Pyrenean mountains, my most steady and communicative correspondent; and therefore it will be my own fault if our epistolary intercourse should languish.

Jack has had a frock of a good dark colour; and shall

have a pair of lamb-skin breeches soon. Jack behaves very well, and is very obliging; and in his readiness to assist, and put an helping hand, often puts me in mind of a gentlewoman that is very nearly related to him. Mr. and Mrs. Etty take a great deal of notice of him, and have him to dine every Sunday.

No doubt your wren that you saw was the shivering wren. So Mr. Lever does not know the grasshopper-lark; that is plain. But as it has done whispering there will be no procuring one 'til next season. He seems also to be unacquainted with the laughing sort.

If you were to recollect you would call to mind that my letters to Mr. Pennant are full of accounts of the sedge-bird. It was by my means that that bird, when omitted totally in the 'Zoology': was inserted in the Appendix.* Mr. Pennant had seen it in Lincolnshire, but did not at all know what to make of it, nor how to ascertain it. He was misled by Ray's classing it among the "*Picis affines.*" † Your appellative of polyglott pleases me so much that I shall adopt it. It has the notes of many birds: and could it be persuaded not to sing in such an hurry would be an elegant songster. They abound with us, especially on the verge of the forest: and are sometimes at James Knight's ponds. In short, wherever there are pools or streams.

Mr. Lever may very probably be right with respect to the short-eared owl. I have always suspected that Mr. P[ennant]'s tawny owl and brown owl were only different sexes of the same species.

I am sorry that you met with such a rebuff at Midsummer, such a cold and dismal summer solstice. For if

* p. 16, plate x. (1770).

† 'Synopsis Methodica Avium,' p. 47, No. 3. In Gilbert White's own copy of Ray's work, now in my possession, he has written "Sedge-bird" on the margin.—A. N.

these things were done in a green tree, what will be done in the dry? Indeed we had at that season cold, wet, black weather: but from July the 7th to this time have enjoyed the most lovely season that ever was seen; both as to sun for our hay, and since soft showers for our meadows, gardens and turneps. I have a prospect of a very fine crop of grapes on the walls of my house. Pray revive your journal without loss of time. The wind was very far from being constantly N. with us for a month before you wrote: it was very much so indeed from June 27th to July 3rd inclusive: but the week before was all S.W. to a day: but then again from June 13th to the 19th it was pretty much N.E. and N.W. at different ends of the week.

I shall write to Mr. Twiss soon, and repeat my invitation. From the time that the widow returned from her bathing in the sea she began to be less cruel; and last week she consented to make Mr. Webb* happy: they kept their wedding at Newton. Mr. Y[alden] was so delighted with the event, that he made verses on it: the strangest verses you ever saw. He made also a copy on his new alcove: they are all alexandrines, and wonderfully unwieldly; and very much like those before 'Pilgrim's Progress.' It is pity that so worthy a man should be troubled with such an infirmity!

The story of the Aurora is all contradicted.

It is pity that you can't hear from Linnæus: you had better write again.

Some boys went to hunt flappers (young wild ducks) last month in the forest: among the ducks they caught some minute wild fowls alive. I examined them, and found them to be young teals; but never had suspected that teals ever bred in our parts 'til now! I redeemed one and turned it into J. Knight's ponds.

* The doctor at Alton.

Thomas Corston has brought Berriman a certificate of his marriage with Mary Gregory in the church of St. George's, Hanover Square, in Feb. 1771. The extract was taken April 24th, 1773. It is on a 5*s.* stamp and looks as if it was genuine. Corston has called since.

This evening I expect from Fyfield brother Thomas and master Brocket; and brother Henry and Thomas Holt White.* The two former are on their way to London.

When did you get into your house?

Jack joins in respects.

Your aff. Brother,

GIL. WHITE.

Harvest does not begin 'til next week.

To-day sweet summer weather after showers.

Bro. Harry was told in Oxford that Linnæus is certainly dead.†

[From Thomas White, on the same sheet as the above.]

I should have sooner replied to your letter on the 11th of last month, but have been much from home since that time; Your wine in bottles is sent to the Care of Messrs Benson and Postlewait at Liverpool directed for Mr. Jos. Fielden, Blackburn I.W. [*sic*]. Soon after I received your letter I fell into company in the Cambridge Stage with Mr. Clapham who is secretary to the bishop of Chester and several others, he says there is but one vicar in the north who has power to let 21 years leases, it is continued by him as an antient custom, but Mr. C. is certain would not be allowed to others, as it is very liable to abuse, altho' it might be beneficial in yours and some other cases. I shall return to London to-morrow and will execute your commissions soon. My Buxton scheme is prevented by business that will not

* Eldest son of Gilbert White's brother Thomas.

† He did not die until 1778.—A. N.

suffer me to be away so long as one might rationally expect benefit from *

I am not surprized at what you mention concerning Mr. P[ennant], it is not improbable but that you may see your birds and fishes of which he has drawings, in some future publications of his. Pray remember me to my sister and believe me to be

Yours very affectionately

THO. WHITE.

To the Rev. John White.

Sep. 11, 1773.

Dear Brother,—Your last letter but one and my last crossed, I believe, on the Road. I am now to thank you for your frank of Aug. 25.

As to Jack he is no trouble or inconvenience to me; but of real use: and therefore I desire he may stay as long as ever you can spare him. Moreover I wish you and my sister, while your house drys, would come and spend one more winter with me. You might put in an elderly, grave person for that period to make fires, and take care of your goods, and defer the hiring of servants 'til spring. I will endeavour to do every thing to make the winter as easy to you as possible: you shall have a bed put up in the drawing-room, and a grate, where you shall have a constant fire, by which you may instruct your son, and fabricate your Fauna. As my sister, I know by agreeable experience, is of an active disposition, she shall, if she pleases, manage my house, and see to provisions; and we shall, if it pleases God to bless us with health, I trust, pass the dead season of the year in no uncomfortable way: and at the return of spring I will let you depart in peace; and will follow you in the summer into Lancashire. All this proposal is the result of a sincere intention: and therefore I hope you will think of

* Letter torn.

it in earnest, and not let the consideration of a long coach-journey (which is not so formidable to either of you as to some others) prevent the comfort and satisfaction I propose from such an undertaking. When you write again I hope to discover that you have considered this matter, and will be so kind as to think of putting it into execution.

Mr. Derham in his 'Physico-theol.' incidentally mentions D[r] Leigh's 'Natural History of Lancashire,' which perhaps Mr. Lever has got, and you should see.—If I err about the *Mot. ficedula,* I err in good company: for your bird is indisputably the bird that Linn. means by *Mot. f.* and yet he all the while acknowledges that it has "mandibula superior utrinque emarginata, et latere vibrissata," which are the characteristics of a *Musicapa.* In Edwards there are cuts of the male and female. As to myself I can't help thinking that your bird is of a *plumper and shorter habit* than the cock coldfinch* with a white forehead: besides you sent many like the cocks that had no white in their foreheads. Mr. Shaw, I see, in the papers, is preferred: I am glad he continues to send you new birds. The lark-like bird may probably be as you say.

As to beer I hop moderately at the making; perhaps half pound to the bushel; and then put in more when it has done working, perhaps half pound to the half hogshead: scalding them in water, and putting them in when cold. They will never sink if put in dry. I have now at tap again excellent beer.—From August 9th to 14th inclusive was most wonderfully hot. My thermometer on the 13th was up to 78½°; on that day in the evening came a violent thunderstorm which about London did great damage. Jack and I were all that week at Mr. Mulso's at Meonstoke. In the night between the 18th and 19th of August there was

* For some reason which I have never been able to discover what is now called the "pied fly-catcher" (the bird here in question) was called coldfinch —a name which seems to be now entirely abandoned.—A. N.

a violent storm from the N. which damaged all our hops to a great degree, and particularly Sir S. Stuart's which were very fine. Hops will be dear, and bad.

Mrs. W. Isaac is with me at present. She brought her son Bap[tist] that he might stand for a scholarship of Winton Coll. we went this week, and left him with the usher, where he is to continue as a commoner 'til taken into College. My wall has produced about 10 dozen of most lovely peaches and nectarines: and I have a fine show for grapes.

I have written to Mr. Twiss, and invited him to come: but he says he cannot possibly be spared from the fortifications 'til Winter.

Swifts left us about the hot week. Young martins continue to come out daily. •

Nanny Woods is quite recovered.

Mr. Etty's Portugal brother is here at present.

How can Mr. Curtis* leave his shop, and go to Gibraltar in pursuit of natural history?

Jack makes English themes, and writes letters: we have gone through Phæd. and now read Virg. Georgics, and Sallust.

Mr. Lever, I find, is an excellent *practical* ornithologist: if we are to correspond, who is to break the ice? and on what subjects are we to exchange thoughts? He will, I hope, study system.

Mr. Bassat's new house is all wild still; no room finished: he will run into great expense.

We have now a very wet and windy season: a sad time for barley-harvest, and hop-picking: most of our wheat was housed in very nice order.

* William Curtis, author of the 'Flora Londinensis,' and several entomological works. He was also the discoverer of the real nature of honey-dew, but not to be confounded with *John* Curtis, a very well-known entomologist.—A. N.

I have some hopes of seeing Skinner soon. Mrs. Isaac, Niece Molly White who came with her, and Jack join in respects.

Y^{r} affect. Brother,

GIL. WHITE.

To Thomas Barker.

Selborne, Sepr. 14, 1773.

Dear Sir,—I can readily give you credit for the change of colour that befell the bulfinch; because when I first undertook the church of Faringdon, the person where I used to dine on a Sunday caught a cock bulfinch in the fields after it had arrived at its full colours. In about a twelve month it began to grow dingy, and losing by degrees its gay apparel, it became leisurely, in I think about three years, as black or blacker than a blackbird, all save some of its wing feathers, which continued white, at least in part. This bird remained in this mourning garb to the day of its death; and lived, I perfectly remember, altogether on hemp-seed; a kind of food which, I have heard before, has a tendency to blacken those birds that live altogether upon it. The owner of the bulfinch had at the same time a skylark which was supported altogether in the same manner, and became very dusky, but not black.

Be pleased to remember that though I happened to have seen a similar case, yet I look upon the phænomenon as odd and extraordinary; and am much obliged to you for your information, and shall be for the future for any curious anecdote that falls in your way.

From the 9th of August to the 14th inclusive the heat was very severe night and day; and on the 13th in the evening arose from the S. that great tempest of thunder and lightening which did so much damage in and about London. The rain attending that storm was of signal service to the hops, which before began to languish. But in the night between the 18th and 19th of August such a wind came from the N. that it well nigh demolished all the plantations. In Sir S. Stuart's garden consisting of 20 acres not one pole was left standing for many acres together; and as his crop was remarkably fine, he suffered the loss of many hundreds of pounds in that one night. Since the storm hops have never thriven; and are now picking; but are small, and brown; and will be very dear and poor and ordinary. My thermometer on the 13th of August was at 78½ within doors.

In the beginning of wheat harvest we had some rain which frightened some farmers and made them house some of their wheat too soon; but on the whole the wheat went in in most curious order. As to the spring corn both here and on the downs it all lies abroad in a bad way, for we have had nothing but rain since Sep. 1. Apples fail in general; I have again, as I had last year, more than my share; but not one pear. My Apricots were almost all cut off in bloom; but I have on my wall about 10 dozen of the best ripened peaches and nectarines that I ever saw, that are now in high perfection. My crop of grapes is very great, of which I shall begin gathering to-morrow; and they will supply my table constantly 'til the frosts strip the trees of their leaves. I wait much on my vines, and have them trained with great care and exactness.

My s^{t}foin was much damaged; but my meadow-hay was got up in nice order. Hay has proved a prodigious crop in all parts. Brother Thomas writes that some farmers in Essex offer to sell it at Xmass next at 25*s.* per ton out of the rick; and a year ago it was with us at £4. 4*s.* 0*d.*

I cannot think that the wheat with us is any thing of a crop.

Your affectionate Brother,

and humble servant,

GIL. WHITE.

To the Rev. John White.

Selborne, Octr 1st, 1773.

Dear Brother,—If my last letter surprized and distressed you, it occasioned very different impressions from what were intended. For therein I proposed to imply how much satisfaction I had received from your company and my sister's last winter: and could not help perswading myself that it might have been in your power to have repeated your visit for one winter more without any great impropriety, and any farther inconvenience than some personal trouble.

You say I may easily throw up my church, and come down to you; which I ought to do, and fully intend: but then when I once relinquish my employ, I cannot reassume it when I please, even though I find myself ever so much becalmed for want of something to call me forth, and employ my body and mind.

I must on the 12th of October set out for Oxford, where I must stay about ten days. There are at College, I find, like to be uneasinesses and disputes about offices, and residence: so that I am not able yet to guess whether I shall not be called on by and by to be stationed there for a time.

As soon as I return Jack and I must set out on a visit to Mrs. Snooke, by whom we have been called on all the summer to turn our horses' heads that way: but something or other has still intervened.

Thus you see my time is cut out 'til the middle of November, and after that my fate is *at present* dubious; however, I will be sure to write to you in a more satisfactory manner from Oxford, at which place I shall hope to hear from you before the 20th, on which day I intend to return.

Pray write *single* on your letters; for the last, tho' only half a sheet, was charged double postage.

Jack joins in respects.

Your affect. Brother,

GIL. WHITE.

Jack has gone through several volumes of the 'Spectator,' and now seems much delighted with Derham's 'Physico-theol.' He does not manage yet to read Virgil with that exactness of quantity that might now be expected; but construes well both in prose and verse. Brother Ben was very urgent with me that you should not become a magistrate without very mature deliberation; because he had been informed that some of your predecessors as magistrates had met with great vexation from the mob of so populous a town. Did he ever mention the matter to you?

To make me consistent with myself you must suppose that I have heard of the probable disputes at College *since* I wrote last to you. However, I do not suppose I shall be called on this winter. We have had nothing but wet, windy, cloudy weather since the first of September. Most of the barley is abroad; some of it has been cut this month. Hops will be dear: they were half destroyed by the wind. Martins and swallows abound.

The statement about the "disputes" at Oriel, confirmed as it is by mention in Mulso's letters of "the perverse party" at the College, rather tends to show that Oxford was much in the same state as when Hearne wrote, in 1726: "There are such differences now in the University of Oxford (hardly one college but where all the members are busied in law business and quarrels not at all relating to the promotion of learning), that good letters decay every day, insomuch that this ordination on Trinity Sunday at Oxford there were no fewer (as I am informed) than fifteen denied orders for insufficiency, which is the more to be noted because our bishops and those employed by them are illiterate men."

ORIEL COLLEGE IN THE EIGHTEENTH CENTURY

[To face p. 228, Vol I.

*From Mr. Skinner.**

Purley, Berks, Oct. 17, 1773.

Dear Sir,—I am extremely obliged to you for the favour of your two last letters; one of which I found upon my arrival at this place some time after the date; the other, after various peregrinations, got at length to Oxford a second time, from whence I have lately received it. I return my sincerest thanks for the very friendly invitation they give me to Selborne; which, from the former specimens I have received there of your obliging hospitality, I could heartily wish to accept. But having been detained much longer in Brecknock-shire than I intended, I am now under a necessity of returning to Oxford in a day or two. However, I flatter myself with the pleasure of seeing you at Selborne some time in the next summer. . . .

In July last I spent an evening with Mr. Banks and Mr. Lightfoot † at Brecknock. They had been botanizing from Bristol through the counties of Glamorgan, Carmarthen, and Pembroke; and had been very successful. In the environs of St. Vincent's Rock only (visited by every botanist) they found three new British plants (one of them the *Arbutus uva-ursi*), and afterwards several others that were either new or very dubious, as not having been found by the botanists of the present age. Among the former, *Sison verticillatum*, a common plant in the meadows of Carmarthenshire; among the latter, *Cheiranthus sinuatus*, *Adiantum capillus-Veneris*, and a species of *Festuca* unknown

* Fellow of C.C.C., Oxford.

† John Lightfoot, born 1735, entered Pembroke College, Oxon., 1753; B.A. 1756; M.A. 1766; librarian and chaplain to the Duchess of Portland; curate of Colnbrook, near Uxbridge; 1772 travelled in Scotland with Pennant, and at his expense published the 'Flora Scotica' in 1778; catalogued the Portland collection on the Duchess's death in 1788, and soon after died himself, at Uxbridge. His herbarium, bought by King George III., is now at Kew. He discovered the reed-wren, and described it in 'Philos. Trans.,' 1785. F.R.S. 1781, and an original F.L.S. His MS. journal of his excursion in Wales is in the Botanical Department of the British Museum —A. N.

to Linnæus, but figured in Ray's 'Synopsis,' Tab. 17, Fig. 2., *Vide* p. 413, with several others I do not immediately recollect. They were bound to Snowdon, in the neighbourhood of which they proposed to spend five or six weeks. I have not heard of their success. But as Lightfoot had with great accuracy collected the names and places of all the curious plants they hoped to find, from Ray and others, I presume the journey, which was new to both of them, has not been thrown away. You have, I believe, seen Lightfoot since his Scotch Tour with Mr. Pennant, and of course have heard that he discovered many plants that had escaped the Scottish botanists. Mr. Yalden is now in Scotland, studying physic at Edinburgh; and, as he is to be part of his time in the Highlands, I make no doubt, will make farther discoveries. He is not expected to return under two years.

About the middle of August I saw our friend Sheffield on a physiological ramble, in a state of uncertainty whether he should cross the Irish Channel or not. I recommended him to Aberystwith partly as the best place in South Wales for a passage to Dublin, and partly because the coast of Cardiganshire was left untouched by Lightfoot and Banks. . . .

The last advice I had of Linnæus was from Mr. Banks, to whom he had written, but a little before, with a good deal of chagrin, for not having sent him the specimens of the South Sea plants he had promised him. He said, "he was an old man; and that if it was intended he should *ever* see them, the sooner they were sent the better." Mr. Banks thought he was well.

I am with great esteem,

Your obliged friend, &c.,

R. Skinner.

P.S.—What advices of or from the Island of St. Matthew? Any new birds from my old friend John White?—or intelligence of the Senegal swallows? I hope they, as well as the Governour &c. have escaped the massacre of the natives, mentioned in the newspapers.

Thos Cooper Bp. of Winton Author of the dictionary, lyes buried in one of the Churches of the Borough of Southwark; I forget which, having neither Wood's 'Athenæ' or 'Godwin de Præsulibus' to consult. The Vicar of Selborne who made the entry seems to have been an oddity.*

From the *Naturalist's Journal*–

"Oct. 21. Saw several martins at Dorchester, in Oxfordshire, round the Church. It is remarkable that the swallow kind appear full as late in the midland counties as in the maritime, a circumstance this more favourable to hiding than migration. As it proved these were the last martins that I saw."

To the Rev. John White.

Selborne, Novr. 2nd, 1773.

I cannot deny that I did receive your letter while at Oxford; but then my short time was so totally taken up with the accounts, and the common room, and a little visiting, that I never had the resolution to sit down to a regular letter.

Just before I came away, Skinner came up from a twelvemonth's sojourning at Brecknock to a regular residence in College; he made me dine and sup and spend the whole day with him; and is the same chatty, communicative, intelligent, gouty, indolent mortal that he used to be. Moreover, he had written me a long letter! just before I saw him, which arrived at Selborne some posts after I got home. He is now first oars with regard to College preferment, and provided Dr. Patten should die (who is in a

* *Vide* 'The Antiquities of Selborne,' Letter VI.

very dangerous way) he will succeed to a living of more than £400 per ann.

Jack did not go with me to College, for I well knew that if he had I could not have been with him one tenth of the time: however, by means of his old friends at the vicarage, and at Newton he was left very little with the servants. We are to set out on Monday next on a visit to Ringmer; and nephew Benjamin who is visiting with his father and mother and sisters at Newton, is to accompany us. We intend to stay two Sundays. Last week I began taking away the earth of my grass-plot and walk at my garden door, so as to destroy the two steps and to level the ground quite on to the alcove. We were much interrupted at first by rain; but now go on swimmingly. When completed the job will be *tanti*, and the nuisance of two slippery steps removed.

Poor Mrs. Snooke writes me word that she has the gout in one hand and both feet, and what is worse a terrible swelled leg that threatens her with the dropsy.

And now as to my visit into Lancashire. I am very desirous, as indeed I ought, to make you a visit. But this cannot be brought about 'til after Xmass; for in the first place our journey to Sussex will take up half the space between this and that; and in the next place I am hampered with a breaking tenant at Greatham, whom I want to remove before I leave these parts. And by Xmass, I fear, the severe weather will much impede our expedition, which will be a sort of migration reversed, to the N. instead of the S. However, if it please God to enable me, I hope to get to you some time in January. I rejoice much to hear that you have at last got possession of your house; and indeed, was I ready now, some time would be needful for you to settle a little before any visitors came upon you.

It is pleasant to hear that the spirit of Natural enquiries subsists still at Gibraltar. There is a muscle, perhaps your dactyl,* that gets into the cliffs on the coasts of Sussex, and terebrates the chalk in a most curious manner. You saw, I think, a piece of chalk so bored at my house. I have received a most violent complimenting letter from Mr. Pennant lately. He is going to publish a second edition of 'British Zoology,' and is to do wonders with the information extracted from my letters. I shall take the opportunity of laying before him the more glaring faults in the first edition.† Ice this morning! My grapes are now delicate. Pray write while I am at Ringmer. Jack joins in respects to my Sister and yourself.

Ring-ouzels in plenty this last Septr.

Y^{rs}. affect.

GIL. WHITE.

On November 29th, 1773, Mulso writes from his Canon's house at Winchester:—

I am booked into strict residence. . . . You who receive Letters and keep up a correspondence with the philosophers of Europe, think little of an epistle from the *Vice-Dean* of Winchester; but my new title and my old friendship are all that I have at present to offer from,

Dear Gil,

Yours sincerely and affectionately,

J. MULSO.

To the Rev. John White.

Ringmer, December 9th, 1773.

Dear Brother,—Jack and I arrived here on Thursday, December 2nd, and found that your letter of November 11th had waited for us more than a fortnight. We were very agreeably surprised to find Mrs. Snooke so much recovered after so great a plunge. She is cheerful and chatty,

* *Pholas dactylus* of Conchology.—A. N.

† That is to say the first *octavo* edition.—A. N.

free from pain, and able to walk to all parts of her house without a stick; rides out in her chaise, and is for her great age an extraordinary woman. Her leg, I suppose, is much swelled; but Mr. Manning does not seem to think that that circumstance is attended with any danger.

In my last, as I wrote on a sunny day, I mentioned that my levelling work went on *swimmingly;* but there came immediately upon me a glut of wet for many weeks that spoiled my metaphor, and drowned and floated all my works, and sadly embarrassed our operations in the clay, which was all converted to mud and mire. Before we could dig we were obliged to lade. However, by perseverance in five weeks instead of twelve days I finished my job, which has a very good effect, though neither the turf nor the pavement can be expected to lie quite so smooth and regular as if all had been moved in dry weather. The ground came out to nothing on an hanging level on the grass plot short of the mulberry tree, and in the broad walk midway between the farther wicket and the alcove on a dead level. I apprehend no harm from the borders; they will be lowered. We went 20 inches deep at the entrance of the long walk. My back front looks higher and better.

I am sorry you have improved your parlor 'til it smokes; it is a common case, but you must exert all your mechanic powers to remove so sad a nuisance. My horse Miller is very lame, and could not come; as Jack was trotting him in the north field he trod in an old track, and strained his knee in a wonderful manner. However, his old friend Mr. Etty has lent Jack his poney, that carries him well. The tortoise went under the ground about November 20th, came out again for one day December 2: and now lies in a swampy border in mud and mire!

My letters to Mr. Barrington swell very fast; he has engaged me in the monography of the swallow genus. I have dispatched the house-martin, and my letter is to be read before the Royal Society. I wish I knew a little more about the bank-martin. Does Mr. Lever know anything about the bank-martin that can be depended on? Harry's little folks, all but Harry Woods, are down with the whooping cough. While Brother Benjamin and family were at Newton, Mr. Pain discovered at London that James White had got a cataract or film on one of his eyes. Brother Harry is going to have a West Indian boy at £100 per annum. If Mr. H. Woods* could be prevailed on to take an apprentice he would be the best Master I know, as he is always at home; but Brother Benjamin asked him the question with respect to Edmund,† but he has never returned any answer. Jack is taller somewhat than Mr. Yalden, and grows large in his limbs. Mrs. Snooke has never heard any more of her lawsuit. Poor Mr. John Warnford, of C. C. C., is, I see, just dead, and his lecture disposed of. If he had a College living, as I take for granted he had, then there is a parsonage at Skinner's option. I fear Mr. W. has left a widow. Easter, I find, comes this year very early, and will cruelly intersect the spring, as I must attend an election at Oriel in Easter week. Your fauna, I fear, begins to languish, as I hear nothing of it in your last. I should be sorry to find that you neglect to arrange and methodise so many fine materials. If you don't make haste I shall publish before you. All friends join in proper respects.

Yrs affect.,

GIL. WHITE.

Returning on December 15th, 16th, by way of Findon and Chilgrove, where his connection Mr. Woods lived, he notes in his *Journal* the exact

* A china merchant, or "chinaman," in the Poultry.

† Son of Benjamin White.

depth of the latter's well, a circumstance which seems always to have attracted his attention. On December 26th, 1773, Mulso writes from Winchester regretting the failure of his efforts to find his friend a *locum tenens* of his Faringdon curacy. He continues—

"Mr. Hinton joined us in abusing you and Mr. Etty, for not making your parish make a road to you, which he averred could be done for very little. Think only of my knowing no time of the year for getting at you *without a guide;* and seldom with one. You are the 'toto divisos orbe Britannos.' Cannot you threaten to vote against Sir Simeon if he will not urge the Thing and help it? What other use is there now in general elections? Have we not here a Duke for our Mayor, and Baronets and Knights for our humble servants, because the Day approaches? If you lose this opportunity, I shall think that you love your Rosamond's Bower, *because* the access is inscrutable."

CHAPTER XI.

To the Rev. John White.

Selborne, Jan. 12, 1774.

Dear Brother,—As I made no manner of doubt but that your many kind and repeated invitations were very sincere, you will, I fear, feel a little disappointed when you come to find the purport of this letter. But I desire you would hear what I have to say before you condemn me.

I wrote to Mr. Roman * signifying a desire of being set at liberty from his cure, and fixing a day. He returned me a very handsome friendly answer, in which he wished me still to continue; and as he understood a desire to visit distant friends made me uneasy under restraint, he was ready at once to advance my salary to £50 per annum,† hoping that sum would enable me to procure assistance from Oxon. or elsewhere, whenever I wanted to take a long journey.

This proposal concurring, as you know, with my constant wish of reserving an employ to return to, has at present put a stop to my N. expedition; and has put me upon trying every expedient for procuring a substitute by writing to J. Mulso, brother H., Skinner, &c., &c. What I could wish is, to be at liberty after Easter to visit you from Oxon. For I must not pretend so totally to disregard College

* Rector of Faringdon, whose curate Gilbert White was.

† It was £40.

concerns as purposely, and in good health, to decline appearing at an election.

For my own feelings I often wish myself with you: and make many comparisons between this and the last winter, not much to the advantage of the present. Last winter I look back upon as one of the most pleasant of my life, when I had my friends about me in a family way; and enjoyed the conversation of relations from whom I had been parted so long. When I do come to see you, you may depend on it I should wish to make you a good long visit: besides the stay that I must make in Rutland.

As to Jack I should wish to have him stay as long as ever you and his mother can spare him. He does not altogether lose his time; because he construes and translates, or looks over maps, or writes letters every day. He has read all the 8 volumes of 'Spectators' through with that relish that showed he understood them: and is much pleased with Derham's 'Physico-theol.' and is now embarked in Brydone's and Sir W. Hamilton's Letters, in which I make him refer all the way to maps. Jack and I want to go to Fyfield soon; and therefore wish to know from my sister whether he has had the whooping cough, which all Harry's boys labour under at present.

My monography on the house-martin is finished, and in the hands of Mr. Barrington, who is so much prejudiced in its favour that he proposes soon to have it read before the Royal Society. Another on the house-swallow is near compleated; from which I propose to proceed to the rest of the British *Hirundines*. Can Lever send me any authentic remarks? Pray tell me over again the story of the swallow building on the dead owl's wings and on the conch, &c. I think I could make a good use of it.* You have done very well, I think, with Mrs. Wodin. Mr. Budd was

* *Vide* 'The Natural History of Selborne,' Letter XVIII. to Barrington.

with me this day to tell me that from this day forward he relinquished all business to Mr. Webb.

The barometer is very irregular of late: we have now this day hard frost with the mercury at 29·2, and our last frost (which lasted from the 30th of December 'til January 5th inclusive) began with the barometer at 29·4 All last Saturday afternoon and night, and all Sunday there was a most extraordinary rain from the N., which perhaps was snow northward. Mr. and Mrs. Yalden were here, and stayed all Saturday night. On Sunday there was such a flood down at the pond, that Thomas and I were over the calves of our legs before we got to Peter Wells's; where we were told that if we proceeded any further our horses would swim. So we returned back: and the rain followed on so the whole day, that I omitted going to my church at all the first time for weather since I have undertaken Faringdon cure. People tell me they have known as big a flood, but never one that lasted so long. On Sunday evening it snowed, and then froze hard. Pray tell me a little about your frosts and weather: what birds do you see? Wild geese, I suppose: have you any wagtails in winter? and what other small birds? You will enliven, I hope, your fauna with some dissertations and an agreeable Journal; and some comparisons between the climates, &c., of Andalusia and Great Britain. Poor old Miller is very lame still. Jack joins in best respects.

Yrs aff.,

Pray write soon. GIL. WHITE.

To the Rev. John White.

Selborne, Feb. 4, 1774.

Dear Brother,—By your writing so very quick your last letter arrived much sooner than I could have expected. Since I wrote to you I have talked with Mr. Robinson the Curate of Colmer, and he informs me that he has not the least doubt but that at the end of the summer he shall be able to take my cure off my hands for some considerable time by the following means. The owner of Colmer living, a young man in the N: is, it seems, to take orders this summer, and to supply his own church. This event will remove Robinson to East Tisted, from whence he can with ease undertake the full duty of Faringdon, and will be glad of the emolument, having a large family. You and my sister will therefore excuse me 'til that season, 'til October I hope, when I shall, God willing, wait on you with great pleasure.

Instead of your making excuses to me, I ought to make many to you for detaining your son so long: but, if you can undergo so much self-denyal, I should wish you could now spare him 'til autumn when we will come down together. He is now of real service to me, and a companion in my solitude. We shall ride down to Fyfield soon; and in March I shall carry him to London. It would be cheaper as well as pleasanter for me to travel this summer than to stay at home; because I seem to be in danger of building. But if I do not stay at this place in the summer, when can I think to enjoy it? and was I at liberty now, Easter would cruelly intersect my time, and spoil all.

Lever is a generous man, and is of vast service to you by lending you all his books: I hope Mr. P[ennant] will bethink himself, and wipe off the imputation of selfishness that he lies under. Mr. Budd has just given his fine harpsichord to brother Harry. Your weather and ours accord very much.

A neighbour carries this to London. So I am in some haste.

Jack joins &c. Y^r^ affect. Bro.

GIL. WHITE.

Berriman has paid Th. Corston the money ordered. I got a bond-paper; but not caring to go to any attorney, drew up as full a receipt on it as I could couch in words. The man seemed staggered at first when he saw he must sign stamp-paper: but complying at last he took the money, and getting without the door cursed us all in a most devout and liberal manner: and so ended the negotiation. B. was so provoked at this usage, that he thinks that he should have thrashed him, had he been well and able. Let me hear soon.

On February 15th, 1774, Mulso wrote again regretting that his efforts to find his friend a temporary substitute had failed. He continues:—

'I am sorry that we shall not meet in London or indeed elsewhere for a good while: for you are called to London, and Oxford. . . . Pray, when you build, let it be a drawing-room upstairs, that you may look on the Hanger; let it be higher than the present, and let it be sashed. 'Monstrous! why this will be a great expence!' True, therefore take two years instead of one to do it in. As you want to decoy your family after you to make Selbourne a place of residence, as well as to enjoy it during your own life, e'en do it in a tempting way. Your brothers will be rich men: and you are yourself the richest man that I know, for you are the only man of my acquaintance that does not want money. Stay, I believe I will except my uncle the Bishop—But I am not so sure of him as I am of you.

"May your *Hirundines*,* as I doubt not, bring in the spring and summer of your fame! I am glad you have entrusted yourself to the Public that you reap your due honours. Jack's 'Fauna' should follow close. . . . I still rave at your roads, which have defeated two or three schemes I had to see you. Airson joined me the other day, and vows he never went to Selborne in his life, but he lost his way."

* The first of Gilbert White's papers on this subject was read before the Royal Society in February, and the second in March of this year (1774). They were subsequently printed in 'Philosophical Transactions,' vol. lxix.

From the *Naturalist's Journal*—

"March 5, 1774. Received as a present from Mr. Hinton (to whom it was sent from Exeter with many more) one of Mr. William Luccombe's new variety of oaks. . . ."

In another hand is written—

"But inquire of your brother of Thames Street about it."

To the Rev. John White.

Faringdon, March 6, 1774.

Dear Brother,—Mr. Webb, as you say, is, I believe, a good natured man, and in good business. I have made some enquiries round about by means of Mr. Yalden;* but the answer was, that at present he did not think it convenient to take an apprentice. So it is needless to say anything farther on that subject.

What is become of your friend General Cornwallis? I have never seen any mention of his arrival.

You never let me know whereabout your expences came for the fitting up of your house. I talk of building a parlor; but when it comes to the point I have fears about the trouble; besides green walls will not be habitable till the second year. In what language did you write to Linnæus? and in what did he answer you?

The late decree in the house of Lords concerning literary property will make booksellers shy of publishing new editions, as it renders such property very precarious. There is in London a strange spirit of decrying Linnæus, which seems to obtain more and more; I think, without any reason. Just as infidels rail at the S.S., though at the same time their minds are much enlightened by them. Mr. Twiss had written me word that he would come and see me in the

* This little incident illustrates the extreme caution which Gilbert White habitually used. Mr. Webb, with whom he was well acquainted, lived at Alton. Mr. Yalden was his neighbour at Newton Valence vicarage.

winter. At Xmas I invited him again; but he was gone into Staffordshire on family business, and when he returned had no more time to spare.

By the death of Mr. John Warnford, Skinner is put in the possession of the living of Bassingham in the county of Lincoln, on the Witham, midway between Newark and Lincoln. His living was set in 1663 at 8*d.* per acre; and so it continues; was it let only at 1*s.* 6*d.* it would be a noble parsonage, now it rents about 200 guineas. Skinner is to stay in college about five weeks longer. If you want to consult him about your fauna why don't you write to him? Direct for Skinner, after five weeks, at Bassingham near Newark.

Jack by no means wants a coat and waistcoat yet; he shall have a pair of fustian breeches. Brother Thomas can hear of no master as yet, but will continue to enquire. Prices are so enormous, that a common seedsman asked him the astonishing deposit of £300! to enable a young man to sell a pennyworth of radish seed!

I could get no *Mus*,* as nobody moved a rick.

Pray write very soon.

Yours affectionately,

GIL. WHITE.

Friends join in respects.

If you have anything for Linn. send it up soon, because the ship is likely to sail shortly. If you have any desiderata with respect to Spain, now is your time, for Mr. Barrington tells me this morning that he has just compleated a nat. treaty with the King of Spain, who is to send all the curiosities of S. America, and the R.S. are to send him all the nat. productions of N. America.

I remember you wished to know how far the *Hirundo*

* The harvest mouse (*Mus messorius*), which he was the first to discover and describe in England. *Vide* 'The Natural History of Selborne,' Letter XII. to Pennant.

melba extended up into Spain, and whether the *Hirundo rupestris* was seen in summer in the internal parts.

Mr. B. shewed me and Jack the curiosities of the R.S.; there was a stuffer packing and preparing the productions of N. America for the use of the King of Spain.

Pray write soon.

During March, 1774, Gilbert White visited his brothers in London. On his return he wrote—

To the Rev. John White.

Selborne, March 29, 1774.

Dear Brother,—The long-contested drawings are lodged in Thames-street in order to be sent down to you. I wish they had been better executed, and the owner* had behaved more like a Gentleman on the occasion. He showed me a letter, which he intended to send you: and as Benjamin and he have literary connections, I hope you will forget, and forgive. They always quarrel and squabble by letter, but accord well when they meet.

You do well to send Linn. your most curious specimens: and not only your most curious ones, but also such common ones as by the circumstances of his country he seems to be unacquainted with. He will, I trust, act with candor; and give you the best information he can concerning your non-descripts. No doubt when he sees the *Pratincola* he will remove it from the *Hirundines.*†

As to my letters they lie in my cupboard very snug. If

* Pennant. The remark is severe, but Gilbert White's letter of 1st August, 1770, shows that when he first sent some specimens from Gibraltar to Pennant, he had distinctly intimated that he should expect to see the drawings, if Pennant had any made by his artist.

† The copy of the twelfth edition of the 'Systema Naturæ,' annotated by Linnæus himself, and now in the possession of the Linnean Society of London, shows that he would have done so had he brought out another edition. He had also marked his account of the species "J. White" by way of memorandum.—A. N.

you will correct them, and assist in the arrangement of my Journal, I will publish. I have finished the monography of all the *Hirundines* except the swift.

Cromall* is so dismally circumstanced that I think there can be no doubt which way I had best act. The late incumbent insolvent, and too negligent to leave any papers of information behind him; no barn; I believe no stable; a wretched house; and all the parish offices for years past in the hands of an attorney, and a gentleman's steward; and to compleat all, the manor-farm belonging to Dr Bosworth's persecuting peer!! Mr. Pen told a man a little before he dyed that he had made one year £160 of his living: but that in general it produced only £150. Dr Bosworth says it may be raised to £200.

Write to Mr. Barrington under cover to the Bishop of Llandaff in London.

I made some enquiries concerning a place for Jack of Mr. Roman. He says the difficulties about settling a young man in London are very great; but recommends an habberdasher, if a proper master could be found. The ribbon-merchant was at Coventry all the while I was in town. Brother Thomas knows a young man in the habberdashery way, and will enquire. He wishes your proposal at Alton could have taken effect.

As usual I had a rash first in town, and then a cold and cough; neither did Jack escape without some little cough. I am feverish and faint with my cold, and must not venture to Oxon. except I mend.

Skinner came to town in his way to Bassingham, and spent half a day in Thames-street. Brother Thomas admired the clearness of his head. A Roman-catholic gentleman makes pretentions to his living: but his claims are looked on as futile.

* The Rectory of Cromhall, Gloucestershire, in the gift of Oriel College.

I have not been at Hawkley yet; but the falling of the fragments from the cliff is the least part of the story: for a slipping below has disordered, and damaged near 100 acres of land, and ruined two houses! The ground is rumpled, and forced up as it were into waves.

I did not suppose your repairs would inform me about new building. I only wished to know what expence would put a good firm shell in comfortable condition. The vexation of workmen, I fear, is great.

I bought in town Dr Campbell's 'Political Survey of Great Britain,'* a work that has employed his thoughts more than 20 years: 2 vol. 4to, £2 2*s.* 0*d.* As far as I have seen I think his stile is not so good as in his 'Present State of Europe.' The new work seems to comprize a variety of knowledge.

Some time ago I put Hume's 'History of England' into Jack's hands. The young man is much taken with the work, and reads it with great earnestness, and in preference to the 'Spectators' and 'Tatlers': and makes pertinent remarks.

With respects to my sister we remain

Yrs afect.

GIL. WHITE.

I found by a Gentleman from your parts that you had none of the floods to the N.W.

The following letter from John White, junr.—a decidedly good one for a boy aged fifteen—fully describes the landslip at Hawkley Hanger, which his uncle remarked upon in an undated letter (and therefore not actually written) to Pennant:—

* The last and most elaborate work of John Campbell, LL.D. (1708–1775), a miscellaneous author.

To Samuel Barker.

April 6, 1774.

Dear Cousin,—At the request of my uncle I intend doing myself the pleasure of giving you some account of an extraordinary event which lately happened in the parish of Hawkley. During the vast rains a large fragment of the Hanger (late my grandfather's)* slipped away for near two hundred yards in length, and fell down the steep to the depth of 40 feet, carrying with it the coppice-wood, hedge, and gate between the two fields, &c. The sinking of this gate is very strange, as it stands at present as upright as it used to do and is as easy to be opened and shut.

The next thing observable is a little hop-garden and pond, in the former of which there are two or three places that are sunk four or five feet: besides many other parts of the garden are very full of large cracks and openings. The bed of the pond is also very much sunk and the place from whence the waste water used to run is now the highest part.

A lane which went down one side of this hill is sunk eight or ten feet, and very much pushed forward so as to be rendered impassable.

There is situated in the same piece of ground with the pond (which is meadow-land) a small cottage, the inhabitants of which were greatly alarmed on the night in which this happened, by a gradual opening of their floor, till at last they perceived that part of the cottage nearest the hills began to sever with very loud noises. The upper part of the cottage is since entirely fallen down.

A neighbouring farm-house is also so much sunk and so full of large cracks as to be rendered not habitable.

There is in one field that was wheat last year pretty well an acre so much sunk, that it is impossible to be ploughed. All the corn land which was affected by this

* John White, J.P.

event is full of large chasms and cracks, some two feet wide: the meadow land has very few of these cracks in it, but seems to be pushed forward; and is filled with large swellings of the turf very much resembling waves: in some places where the ground met with any thing that resisted, it rose up many feet above its former surface.

In one place four or five trees are driven all together in a huddle. One tree is entirely beat down.

It is supposed eighty or an hundred acres of ground are damaged by this accident.

There has been a great concourse of people to see this Event. It is computed that a Sunday or two after it happened there were pretty near a thousand from different parts of the country. One of the persons to whom the cottage belonged has lately been about with a petition in order to attempt to rebuild it. Hoping that this account of mine will give you some idea of this wonderful accident; with duty to my uncle and aunt and love to my cousins, I remain

Your affectionate cousin,

JOHN WHITE.

To the same from Gilbert White (on the same sheet).

April 6th, 1774.

Dear Sam.—Molly White* has left school, and is now, I believe, in London; towards the end of this month her father brings her down to this place, intending to put her for a while under the care of Mrs. Etty. Poor farmer Berriman has been for some weeks in St. Bartholomew's hospital for a terrible disorder, which he concluded to be a stone in his bladder; but on Monday last he was re-

* Mary, the only daughter of Thomas White, now 15 years of age. She used often to stay at Selborne, and was the "little girl" mentioned in 'The Natural History of Selborne,' Letter LIX. to Barrington, as remarking that the rooks when flying round and cawing at dusk were "saying their prayers," as her brother Thomas notes in his copy of his uncle's book.

turned as incurable: the surgeons suppose his case to be an ulcer in his bladder. Boxal the blacksmith is dead, and his widow is removed to live with one of his daughters. His house and shop sold for a great price. These last sentences are addressed to your mamma. Nanny Woods continues remarkably stout; but her poor father is in a very poor state of health, overwhelmed with business, and neither able to go on, nor willing to relinquish. Jack and I are newly returned from London, where I caught a great cold. Tell your papa and mamma that I hope they will please to come and see me this summer; and will bring you, and as many of your sisters as is convenient. Among other things you will be glad to see the strange sight described above. I have been prevented as yet by indisposition from seeing it myself. Captain Cook and Mr. Forster, it is expected, will be at the Cape next November and home about next March. The S. and W. of England have suffered lately in a wonderful manner by floods; but I found by a gentleman who arrived in town from N. Wales in the midst of all those bad doings, that nothing extraordinary had happened in that way on the N.W. side of the Kingdom; and so I find by my brother John's letters. The landsprings, or lavants, are higher on the Hants and Wilts downs than ever they were known in the memory of man; and so they are at Faringdon. Your affectionate Uncle,

Pray write soon. GIL. WHITE.

To the Rev. John White. Selborne, April 29th, 1774.

Dear Brother,—After thanking you for your letter of the 12th, I should inform you first (as it is matter of Business) that it will not be in my power to pay Mr. Willis,* because

* Of Holiburn School, near Alton, to which John White, junr., had been sent in 1770, or earlier. He was, however, at present under the tuition of his uncle, to whom he became a useful amanuensis. A good deal of the MS. from which the 'Selborne' was printed is in his handwriting.

I laid out much more in town than I expected, and am bare of cash.

Out of all my journals I think I might collect matter enough, and such a series of incidents as might pretty well comprehend the Natural History of this district, especially as to the ornithological part; and I have moreover half a century of letters on the same subject, most of them very long; all which together (were they thought worthy to be seen) might make up a moderate volume. To these might be added some circumstances of the country, its most curious plants, its few antiquities; all which together might soon be moulded into a work, had I resolution and spirits enough to set about it.

As to your own work, your journal incidents will be the most entertaining part of it. Skinner was much pleased with them. You should moreover, I think, have some letters or dissertations for the unsystematic part of your readers, who will not so well relish a Fauna alone. A comparative account of the climates (where they will admit) would, I think, be very engaging. You have by you an abundance of your own letters.

I should have told you before, that I have finished my monography on the British *Hirundines.* If you can procure yourself a frank, and will send it to me, Jack shall transcribe you one sheet, that you may see how I have acquitted myself, and whether you approve of the manner.

My cold and feverish complaint hung so about me at Easter that I did not go at all to the election, expecting every day to be laid up; but I wrote to the Provost, and sent a renunciation of the living of Cromall, as I have done since of the living of Swainswick near Bath, which Pen dyed possessed of.

Brother Thomas and Molly will be here, I think, next week in order that Molly may be settled for a while under the care of Mrs. Etty.

As to building I have not yet absolutely determined about it; though next week will be the last week of asking; because the bark of oaks (as I must cut some timber) will not strip or run after that time. I am quite at a loss about the house-martins, as they do not appear yet: the swallows and bank-martins were early: a pair of swifts came to the church yesterday.

Mr. Lever, I find, continues wonderfully generous, and helpful by means of his books: if it was not for the General [Cornwallis] you should, I think, dedicate your work to him. In my monography I have made honourable mention of his museum. Pray where did he get his little mouse? I mean by that to enquire whether it is common in Lancashire. For I concluded, as it had escaped Ray, that it did not extend far northward.

When Brother Thomas comes I will enquire what farther information he has procured with regard to a Master for Jack. It is a very difficult matter, it seems, to find a place where all things concur to our wishes. Jack grows stout and tall, but not upright. He says he is 5 feet six inches and a half without his shoes. We began to read Horace's Odes together: but found many of them so indecent for a young man, that we have taken to the Epistles, which are a fine body of ethics, and very entertaining, and sensible. Jack construes well; but makes slips still with regard to quantity. He is become a very good transcriber. I believe such employ now and then is not bad for a young man, since it teaches him the use and power of stopping; a thing much neglected. Jack has read all the Georgics through.

I wish I could read Reaumur and De Geer.

The spring is forward, and we have had some lovely weather; but winter is returned to-day. I want to know what Linn. will say to the supposed *Junco* and the *Turdus*.

My wall-trees promise much fruit. We join in respects to my sister. Pray write soon. Yr affect. Brother,

GIL. WHITE.

From the *Naturalist's Journal*—

"May 7th. In some former years, I see, house-martins have not appeared 'til the beginning of May: the case was the same this year: and yet they afterwards abounded. These long delays are more in favour of migration than of a torpid state.

"May 24th. *Ophrys nidus avis*, Brother Thomas. This curious plant was found in bloom in the long Lythe among the dead leaves under the thickest beeches, and also among some bushes at Dorton."

The following letter is from Sampson Newbery, Fellow of Exeter, who was apparently the "observing Devonshire gentleman" mentioned in Letter XXXI. to Pennant, as giving Gilbert White information about the breeding and migration of ring-ousels in that county. His father, the Rev. Sampson Newbery, was of Zele, in South Tawton, Devonshire.

From S. Newbery. Exeter College, 14th June, 1774.

Dear Sir,—Both your letters I have been favoured with; one while I was with my father in Devonshire; the other I found on my table at my return to college last Sunday. . . . Whether my engagements here will permit me to make an excursion into Hants this summer, I can't yet easily determine: of this you may be assured; that whenever an opportunity presents itself, an inclination shall not be wanting on my part.

My acquaintance with birds and their history is so slender, that I fear all the information I can furnish on that subject will afford you but small satisfation. Such answers to your enquiries as my observation will supply I shall very readily communicate.

The ring-ousels, which breed about the sides of small brooks upon Dartmoor, and continue there till the end of September or beginning of October appeared there this spring before the end of March: If they are so early every year (and to the best of my Remembrance they used to be so early) may we not suppose that these that visit you in spring and autumn are only the scattered ones that accident has separated from their companions? or perhaps part of a second flight in their way to or return from the more northern part of our island. Whence they come when they visit us in the spring, I confess myself ignorant: but that they cross the British Channel seems to me more probable, than that they should leave us to go either to the north of England or north of Europe during winter, and return to us with the spring. I can venture to assert that the grey crows never breed upon our moors: I never saw, or heard of one seen there; but I once, and once only, saw a considerable number of them together near the south-west coast of Devonshire some time in the winter. This spring the *Hirundines* were very late with us; the first swallow I heard of was seen on the 25th April, and on 1st May we had the swifts plenty; but the house-martin was not seen with us till the middle of May, which is very extraordinary. The bank-martin as well as the nightingale is seldom seen in Devon; but the red-start, white-throat, and fly-catcher are pretty common there: the blackcap I never saw.—Why the redwing and fieldfare should leave us at the approach of summer, and what becomes of them at that season, I have often wondered; in Devon they are seldom seen after March; but I have often seen small companies of them in Oxfordshire towards the latter end of April; and once in the month of May; yet I never heard of a nest of either of these birds. The Stone-plover is a bird I am totally unacquainted with: if it be ever found in Devon, it must be about Exmoor in the northern part of the County. Of the

Scolopaces we have two, the Curlew and the Common Snipe, which breed upon the boggy parts of our moors. The first of these I saw this season March 30th, the other is with us all the year: the Jack snipe too I have reason to think, from the numbers I saw of them in April last, breeds in the same places with the two former. The Whistling-plover breeds upon some of the largest and most retired bogs in the extensive forest of Dartmoor; and about some of our little rivulets we have a little bird of the *Scolopax* tribe which now and then breeds with us; it is known among the common people by the name of Sanderlin;* but whether it be different from the stint or no I can't tell, having never seen any of them in hand. The turtle is a scarce bird in Devonshire; and seen there, as far as I can learn, only in the summer. We have one bird, which, though a scarce one, take England through, is often met with about our brooks: I mean the water-ousel; it commonly builds under the arches, or in the piers of old bridges.

The Starling, which breeds in Oxon., Berks, Wilts, Dorset, and, I suppose, Hants, is never seen, that I know of, in Devon during summer; pray, sir, can you inform me what should make the last county more unsuitable to this bird in summer than the five before-mentioned counties?

A rare Plant of the Orchis family, *Limodorum austriacum* of Ray, is said to grow near Alton. If you could procure a specimen or two for me you will greatly oblige me, who am with much esteem

Y[r] humble servant,

S. NEWBERY.

* It was figured by Albin under this name in 1738, but is what is now commonly known as the summer-snipe or common sandpiper, the name sanderling being now appropriated to a very different bird.—A. N.

To the Rev. John White. Selborne, June 18th, 1774.

Dear Brother,—It pleases me to find that my little monography entertained you; it makes its own excuse for being short. The rest will not offend in that way. Pray, make *all* the *objections* freely that occur.

Linnæus's letter is polite and entertaining and instructive. But, pray, what does he mean by saying that your *Hirundo hyemalis* (for so I shall still call it) is *varietas apus*? for the *apus* and *melba* only perhaps have omnibus *quatuor* digitis anticis; while your swallow has a back-toe like other birds; besides the bill of your *apus* and *melba* are much *bent*, but that of the *Hirundo hyemalis* is *streight.*

I have just received a letter from Mr. Pennant, wherein he says that he has lately sent twenty-nine more drawings to Brother Benjamin, being the whole that were copyed from your Gibraltar cargoes. They are to be kept as long as wanted.

Hambledon* is a place that I have a strong dislike to, on account of its morals and dissipation; besides Mr. Hale has a young partner, should he want to leave business by and by.

My anecdote from Valentia is taken from Willughby's 'Ornithology.' In one of your letters you regret the trouble of transcribing your Fauna. A writing master would take this trouble off your hands for a small sum, but with this disadvantage, that no man can transcribe his own works without seeing plans that he can alter for the better; a benefit which is entirely lost where a stranger is amanuensis. I wish Jack would earn your book; I mentioned the conditions, at which he smiled.

Pray return the printed monography without fail. If

* A few miles from Bishop's Waltham in Hants. When curate of Durley Gilbert White would have known it well. In 1753 his account-book shows a visit there to Mr. Hale, who was probably a surgeon about whom John White had now inquired on his son's behalf.

Mr. Lever desires to see my papers, let him have them; he is a man of honor and will not suffer them to be transcribed. I beg his objections and additions. After I have heard from you again I will send the fourth monography with Mr. Willis's receipt, &c.

Linnæus still allows your *Piscis thoracicus* to be a new genus; he should give it a name, and you an engraving. When he talks of making honourable mention of you in his future edition you may reply "orna me."

No doubt the 'Voyage to the Hebrides'* is an entertaining work; I have seen the cuts, but not the book. You should observe when your summer birds come and depart, that we may compare notes.

It is a sickly time among young folks, especially in Sussex. Thomas has had a pleurisy, and is not recovered yet so as to do any hard work. Molly White likes Selborne and looks well. Nanny Woods is very stout and very brown; she is all day in the sun. Wall fruit abounds, and vines promise to blow well. Harry has just lost his best cart-horse; he is most unfortunate in horse flesh.

As my little mouse is so common in your parts and probably in all the intermediate, I am amazed that it should so long remain a nondescript.

Printers, I know, some of them get a deal of money; but their state of apprenticeship must be, I should suppose very low and servile and unwholesome; but of all this I am not sure. Chemistry, Brother Thomas says, is very unwholesome. A packer's was, I have heard, a fine business, but at present they say merchants are their own packers. I have heard no more of the Ribbon man. Jack will by no means obstruct my motions this summer, but is of use and service. We never fall out much and then it is chiefly about *quantity* in verse, and there is no moral turpitude

* Pennant's 'Voyage to the Hebrides' was first published in this year (1774).

in long and short feet. As to law I have nothing to say about it; lawyers get all the money. Our father* you know did not approve of it. Sure London is so large a field, and you have so many friends there, that you cannot be long without hearing somewhat for your son. That you may get a proper place to your mind I heartily wish.

Y[r] affectionate Brother,

GIL. WHITE.

We join in respects to my sister. Pray write soon. I do not know that our Coll.† is Dutch.

On July 11th, 1774, Mulso writes from Meonstoke :—

"You desired me to let you know when we were peacefully set down at Meonstoke. . . . I have two great griefs: one that I cannot ride; the other that you are accessible by no other vehicle. This last is the highest reflexion upon you and Mr. Etty: God forgive you. I talked of you lately with my neighbour Airson at St. Cross; we abused you on this subject. You ought to love us for it, for I should not care a halfpenny about the road to Selborne, if I had not a regard to Etty and a love for you."

To the Rev. John White. Selborne, July 15, 1774.

Dear Brother,—Jack and I went down to Fyfield on June 22nd, and stayed there 'til July 8th and found and left brother Harry very brisk and very busy. For what between seven boarders, six children, some farming, three churches, and some building, he has enough to do. He has a one hundred-pound boy in the room of Mr. Brocket, who by this

* Himself a barrister.

† This letter is printed from a copy in a female handwriting. "Coll." seems to have no meaning here, and is very possibly a mistake for some other word.

time, I fear, is dead; and Jack and Jem* from Fleet street. The house, especially the kitchen, is so small for twenty-two in family, that Harry is embarked in building a wing of thirty-six feet from the kitchen towards the brew-house; this wing is to consist of a staircase, a large kitchen, a parlor over, and a large garret. The kitchen is to communicate with the new staircase of the brew-house by a narrow entry running against the chalk-wall; and thus the whole house may be made use of without a person's going abroad. The chambers over the brew-house are finished, and very neat. All these buildings will cost a great sum, and prevent the laying up of money; but as more room is absolutely necessary, and things are all done in the way of trade, I hope they will answer in the end.

I am glad to hear Harry talk didactically about grammar rules, and moods and tenses; for unless a school-master is somewhat of a pedant, and a little sufficient in his way, he must expect to be soon jaded with his drudgery.†

Swifts, as I suspected, invariably lay but *two* eggs; and as they breed but *once*, their encrease is very small! I got Harry's bricklayer one evening to open the tiles of his brew-house, under which were several nests containing only *two* squab young apiece; and moreover his workmen all told me that, when boys, they had invariably found only two eggs or two birds. If I lived at Fyfield I should be more learned in swifts; for as you sit in the parlor, you see their proceedings at the brew-house.

I thank you for your strictures on my printed monography, and wish you would extend them to the rest. I *had* used the pronoun personal feminine to my swallows; but somebody objected, so I put *it* in its place; but I think your are right, and shall replace *she* and *her;* though there

* Sons of Benjamin White.

† A remark which singularly indicates its author's shrewdness.

is this objection, that in a *pair* a male is implied, as well as a female; and yet *he* would sound awkward for a bird.

Linn. is wrong past all doubt with respect to your *Hirundo hyemalis;* the *melba* is undoubtedly *affinis* to the swift, but not your new swallow; it can by no means be a *varietas apus.*

The rest of Mr. Pennant's drawings are here; you shall be sure to have them as soon as I can find a safe method of conveyance. I am glad your Fauna swells so fast. You must publish a quarto work; every man now publishes in quarto.

It is not to be wondered at that you do not readily find a master for Jack; for our brother Benjamin, you see, is under the same difficulty with respect to Edmund.

Jack's epistle is of his own composing; I have altered here and there a word, but inserted no sentence.

"Orna me," is a sentence very familiar to me; it is, I think, in Tully's letters.

Mrs. Barker writes me word this week that now she and Mr. B. are come to a determination not to stir from home this summer. They plead the ill health of old Mrs. Barker; and a great deal of repairs to be done to a farm-house.

We have had continual showers for three weeks past; much hay, especially clover, is totally spoiled; and much meadow-grass, and among the rest my own, is still standing, but growing too old.

Harry still preserves your Barbary breed of fowls distinct; but they labour under a disadvantage not very convenient for this climate; for their down comes off before their feathers appear; and so they are stark naked for a fortnight, a condition not very suitable to a cold, wet summer. They are taller and more erect than our breed.

As to franks pray think no more about them; for unless anything particular wants to be enclosed I never wish to receive one again; there is no money that I pay with more satisfaction than postage of letters from friends.

If you will object to the rest of my monographies where you see occasion, I shall be glad.

My neighbour Mr. Robertson is disposed, and thinks he shall be able to take my curacy in the autumn; when I much wish to see Lancashire: but then it must not be 'til the beginning of November, for I must be at Oxford in October, and I am embarrassed with a broken tenant at Greatham, by whom I shall lose a year's rent if I do not look about me at Michaelmas.

Swifts breed but once in a summer, and but *two* at a time; swallows and martins breed twice in a season, and from *four* to *six* at a time; so that the latter encrease at an average five times as fast as the former.

With respects to my sister I remain
Your affect. brother, GIL. WHITE.

To the Rev. John White. Selborne, Aug. 11, 1774.

Dear Brother,—My best thanks are due for your strictures on my monographies, of which I shall avail myself in many particulars, and hope you will extend them to the swift. But I am greatly disappointed with respect to Mr. L[ever] from whom, as a good practical ornithologist, I expected several new remarks and observations, but I don't find that he sent you *one.* Surely that gentleman's scheme with regard to his museum is a strange one; for as I cannot suppose that a man of his spirit will take money, so, if he entertains that great beast of a town for nothing (gratis), it will cost him thousands and be quite a ruinous expence! Swifts may breed twice at Gibraltar, as far as I know, because they arrive there much earlier than with us. But it is past all doubt that with us the case is quite different; for they do not arrive 'till the first week in May, do not hatch 'til the middle of June, and this year and the last departed by the seventh of August. Down at Fyfield I opened the eaves of some roofs on June 30, and found squab helpless young, and invariably but two in a nest; so that their congeners, who lay all round from four to six eggs, and produce invariably two broods in a year, increase five times as fast as they do. Linnæus is wrong even in doubting whether the *Hirundo hyemalis* be a variety of the *apus;* for the feet, the bill, the whole habit bespeak it an absolute swallow, not a swift. The *melba* is an absolute *apus.* Pray send me a copy of that gent.'s last letter; I love to see the letters of great and able men in any way. Don't regard the want of franks. I pay 2½d. for a nonsensical newspaper, and shall I hesitate to pay 7d. for the sight of an epistle from the greatest naturalist in Europe? He will insert you in the next edition of his 'Syst. Nat.': say to him in your next, "orna me."

I cannot procure a grass-hopper lark. They are such shy, skulking varlets; such troglodytes, such hedge-creepers, there is no knowing where to have them. When a good opportunity offers, I will send the drawings to town. Poor Nanny White of Fleet Street is in a very declining way. Happening to meet Mr. Edm. Woods lately at Newton, I pressed him very earnestly to lend his assistance in looking out a master for Jack. He promises to do his best, and is a likely man to forward such a business. . . . You must read Lord Chesterfield's letters; they are very entertaining, though in parts very exceptionable.

Bear is a sort of barley: Mr. Pennant should have told his readers as much in a note. Mr. and Mrs. Lort are now on a tour round N. Wales. Mr. Leach has lost his cause with respect to the lead-mines, and stands obliged to refund all his gains. Molly White* must lose her share, of course. Niece Molly joins in respects. Your loving brother,

GIL. WHITE.

I am mending my tiling and pointing against winter. A stitch in my side makes writing irksome.

* Her mother's maiden name was Leach, of Loveston, near Pembroke.

From the Rev. J. White.

Blackburne [probably August, 1774].

Dear Brother,—I have sent brother Benjamin word of what Linn. says about his new edition.* This roundabout method will dispatch nothing. Brother B. must write his own proposals, if he means to deal with him at all. Surely you and I could manage to correct the press; at least in the Zoological part; the Botanical I must not pretend to.

Have you any queries to ask Linn. before I write again? You see he is willing to communicate, though busy enough.

This new enterprize of my friend Lever's disturbs me on many accounts.† In the first place, I wished *you* to see the museum in its native spot, Alkrington; in the next it will be gone from this country, of which it was one of the greatest ornaments; and *thirdly* it will rob me, I fear, of my friendly neighbour for a great part of the year. I heard from him immediately on his arrival in London. His plan is, he says, "to pursue Natural History and carry the exhibition of it to such a height as no one can imagine; and to make it the most wonderful sight in the world."

Upon this plan I think he is right to exhibit in London, where he will not only collect with more speed, but also make the thing defray its own expences, which no private fortune alone could possibly equal. If you can send the drawings to town while Mr. Lever is there, he will bring them. I shall go to see him as he returns.

What leisure time I have I employ in collecting insects, which I have promised Mr. Barrington as a beginning of his 'Fauna Britannica.' I wish Jacky would pick up all the

* This seems to have enclosed Linnæus's letter to John White of July 3rd, 1774, the last the latter ever received from him, in which the Swedish naturalist mentions that he hopes to bring out a new edition of his 'Systema Naturæ' in the coming autumn. "Si tuus frater edat, certus sum quod hoc prodest optimis typis, qui anglis communes."—A. N.

† Mr. (afterwards Sir Ashton) Lever about this time moved his museum to London.

variety that he can, put them into spirits, and there let them remain till I can get them. Pray examine your sands for the *Myrmeleon.* If in England I know no likelier place. Why should not England have it as well as Sweden? I now recollect that I promised some remarks on your Swift in my next. I have a few observations to make on that bird, but no criticism on your dissertation; and therefore I thought you would be better pleased to see Linnæus's letter.

I am drawing towards the conclusion of my insects; and shall then proceed to the quadrupeds, birds, and fishes. After all there must be a general correction and transcript of the whole, which will be no small undertaking.

We have had a sad, gloomy, wet, chilly season. We are now sitting over a fire. I have brushed up my house as spruce as if it were for sale; but it is to give you as agreeable an idea of Lancashire as I can.

Mrs. White is well, and joins in best wishes and respects with, dear brother, Your most obliged and affect.

J. W.

CHAPTER XII.

To the Rev. John White.

[With a copy of verses by S. Barker.]

Selborne, Sept. 26th, 1774.

Dear Brother,—Some years ago, when I met Sam Barker at Ringmer, I found he had some propensity to poetry, though it had not then been called forth. I therefore then gave him a few instructions; and this summer sent him some old verses on Selborne new furbished up.* These small encouragements have occasioned the lines above, which Jack transcribes for your amusement. By all means recollect the specific difference, or else get new specimens of the *Sphex formicarum falco;* for it is a great pity that so diverting an account as that which you give of that insect should be lost to your Fauna.

You must be sure to caution Linn. that the *Œstrus curvicauda* is by no means the parent of the Star-tailed water-maggot, tho' Derham says it is. Swammerdam, Geoffroy, and others have since discovered that wonderful maggot to be the offspring of a *Musca hydroleon.* See Geoffroy—he calls it *Stratiomys.* You will be the means, I perceive, of correcting many mistakes, and new arranging many misplaced articles in 'Syst. Nat.' which I wish to

* The 'Invitation to Selborne,' which was several times altered and rewritten for various friends.

see as perfect as possible. I wish you may catch a *curvicauda* before the autumn is over; you will at once be convinced that it is an *Œstrus—os nullum, punctis tribus,* &c. It sometimes haunts upland fields, and teazes the horses at plow; but is more frequently found in swampy wet places, and is probably aquatic in its larva state. The nits laid on my horses at Meon-stoke Aug. 19th stick on their legs and flanks still. Skinner can, if he pleases, send several queries and pertinent enquiries relative to 'Syst. Nat.' which he well understands with its comparative merits, and defects, as far as the Author has borrowed from or imitated Mr. Ray. Next summer I will if possible get a grasshopper lark; but they are not easily procured: they skulk in the hedges like mice.

When you write to Linn. next, pray talk to him about tortoises. There are tortoises whose shells are always open *behind* and *before* "apertura testæ anterior," as he says himself, "pro capite et brachiis; posterior pro caudâ et femoribus." These apertures are supported, as it were, by pillars on each side and can *never* be closed. Of such construction is the shell of Mrs. Snooke's present *living* tortoise, Timothy. But then there are tortoises whose under shell has a *cardo,* an hinge, about the middle of their bellies, commanding one lid or flap forward, and one lid backward (like the double lidded snuff-boxes) which when shut conceal the head and legs and tail of the reptile entirely, and keep out all annoyances. Two such (very small they were) Mrs. Snooke had formerly; and the shells lie still in her room over the hall. Now concerning shells of this construction Linn. makes no mention at all; and this construction is certainly the most *curious* and perhaps the most uncommon. But what I would infer from all this is, that in his genus of *Testudo* express mention should be made of this diversity; and the genus should be divided into *testis clausis, cardinatis*; and *testis apertis,* or by some such expressions. He loves, you know, to subdivide his long genera; and such subdivisions are very consistent with system. I hope I have expressed myself so as to be understood in this matter; and I wish you would mention it to him, and ask if such a subdivision would not be proper.* Thus he divides his genus of *Anas* into "rostro basi gibbo," "rostro basi æquali," &c., &c. Can't you hear yet where and how the *Panorpa coa* feeds? It is a pity that part of its history should escape you. I shall transcribe my Swift anew before it be read (redde) to the R. S.—I have alterations of some consequence to make, such as the number of its eggs, &c. It is remarkable that none of the Martins round my buildings have *this* year any second broods. Is this owing to the cold, wet, uncomfortable equinox? We had sweet weather this August 'til towards the end: but since sad doings; much corn damaged and our fruit spoiled. I fear much for Mr. L.[ever?] not loving enterprise and adventure. Poor Nanny White lies now at Newton in a dying way: she has lost all her flesh and all her appetite. Edm. Woods is your only man; if a settlement is to be found, he, I should think, will look it out. We have vast quantities of Hops, but they are mostly distempered. My thanks are due for your kind letter of September 11th. Some of the contents are very strange! Over Leicester house gate must be painted in letters of gold *Museum Leveriense.* Jack's leg does not swell at all of late. I am sorry for your accident, having experienced myself somewhat of the same sort, on the same occasion, but not in the same degree.

Our respects to my Sister,

Y^rs^ affect.

GIL. WHITE.

Berriman lies in the same sad way.

* This distinction between the species of tortoises was proposed to Linnæus by John White in a letter dated October 8th, 1774. He did not however mention it as being his brother's suggestion.

Mr. B. thinks your lameness is the cramp.

Just as I had finished thus far your entertaining letter of Sept. 13th came to hand. I am convinced by Reaumur with respect to the *Œstrus bovis*, &c., and much pleased that you have discovered and seen with your own eyes, and caught with your own hands a *curvicauda*; but it seems strange that you did not see eggs on the horse's hair! perhaps the color of the horse might prevent your seeing them. Horses of dark color are quite discolored by them. I wish to know the *nidus* of the great *Tabanus* also; it haunts, I know, watery, moory places. The *Sturnus collaris* can't be a *Fringilla*, since it has no conic bill. All grass-horses now in watery places have nits. I am pleased to find that Insect extends to your distant part. You may see with your own eyes the parent deposit each single egg. Your remark concerning the bivalve shape of the egg is a good one.

To the Rev John White.

Oxford, Octobr. [1774.]

Dear Brother,—Your letter of Oct. 2nd arrived just as I was prepared to set out for this place, to which I was called a day or two sooner than I intended on account of the University election. I left brother Thomas behind at my own house; and brother Ben. at Newton, intending to club for a post-chaise in a day or two, and to return to town. Brother Thomas has been bathing in the salt-water at Lymington for rheumatic complaints. Poor Nanny White has rather been better for a day or two past, and has rested, and slept a little, and shewed a little appetite; so that her friends were willing to flatter themselves that her illness might take a better turn. She never had any cough.

I heartily wish it was as much in my power as in my inclination, to assist you in the concern you mention. As to Mr. Hill, I never heard of him but at Fyfield, and therefore can only echo my brother Harry with regard to his business, and reputation. He has, it seems, a partner, concerning whom it will be as needful to enquire, as about the principal. There is also a Mr. Baverstock at Marlboro' (one of the Baverstocks of Alton) a man in a flourishing way, with whom brother Ben. is acquainted, who should be asked concerning the circumstances, temper, &c. of Mr. Hill; and the common, trite observation that there is somewhat of adventure, and hazzard where a man strikes out into many businesses should not be totally disregarded. The variety and extent of the business must moreover occasion the frequent absence of the master, and subject his people to be left pretty much to themselves. These matters I have thrown together as they occur, though no doubt they have all been considered well by you before.

As to the business of my journey I have carryed it continually in my mind, and have been still labouring the point: as to my neighbour Robertson of East Tisted, he is very *willing* and *desirous* to help me; but then he has it not in his power without he can find somebody to take one of his churches off his hands. He has met lately with a person that wavers about it, and will send him a final answer soon. Upon this contingency at present does my Lancashire visit seem to hang.

Pray write soon. I return hence on Wednesday Oct. 19th. Respects to my sister.

Your affect. Brother,

GIL. WHITE.

Fine weather for a week after astonishing rains.

On October 20th, 1774, Mulso writes :—

"I have been very much chagrined that I could not hook in a visit to you. . . . I hope you and yours and your good neighbours are well. I shall be glad to hear from you at Winchester, and to see you; you cannot be so displeased at our not coming as we are grieved at not visiting the old and dear scenes, and my old friend their master."

The following letter to his nephew exhibits its author in the character of an amiable uncle, though his conception of poetry may now perhaps seem somewhat strange and out of date. The last paragraph contains a remarkable testimony to his brother John's ability as a critic and writer of verse. Nothing of this is known to be now extant.

To Samuel Barker.

[With a copy of the 'Invitation to Selborne.']

Selborne, Nov. 3, 1774.

Dear Sam,—When I sat down to write to you in verse, my whole design was to show you how easy a thing it might be with a little care for a nephew to excell his uncle in the business of versification; but as you have so fully answered that intent by your late excellent lines, you must for the future excuse my replying in the same way, and make some allowance for the difference of ages.

However, when at any time you find your muse propitious, I shall always rejoice to see a copy of your performance, and shall be ready to commend, and, what is more rare and more sincere, even to object and criticise where there is occasion.

A little turn for English poetry is no doubt a pretty accomplishment for a young gentleman, and will not only enable him the better to read and relish our best poets, but will, like dancing to the body, have an happy influence even upon his prose compositions. Our best poets have been our best prose writers; of this assertion Dryden and Pope are notorious instances. It would be in vain to think of saying much here on the art of versification; instead of the narrow limits of a letter, such a subject would require a large volume. However, I may say in a few words that the way to excell is to copy only from our best writers. The great grace of poetry consists in a perpetual variation of your cadences: if possible no two lines following ought to have their pause at the same feet.

Another beauty should not be passed over; and that is, the art of throwing the sense and power into the third line, which adds a dignity and freedom to your expressions. Dryden introduced this practice, and carried it to great perfection; but his successor, Pope, by his over exactness, corrected away that noble liberty, and almost reduced every sentence within the narrow bounds of a couplet. Alliteration, or the art of introducing words beginning with the same letter in the same or following line, has also a fine effect when managed with discretion. Dryden and Pope practised this art with wonderful success. As, for example, where you say "the polished beetle," the epithet "burnished" would be better for the reason above. But then you must avoid affectation in this case, and let the alliteration slide in, as it were, without design; and this secret will make your lines bold and nervous. There are also in poetry allusions, similes, and a thousand nameless graces, the efficacy of which nothing can make you sensible of but the careful reading of our best poets, and a nice and judicious application of their beauties. I need not add that you should be careful to seem not to take any pains about your rhimes; they should fall in, as it were, of themselves. Our old poets laboured as much formerly to lug in two rhiming words as a butcher does to drag an ox to be slaughtered; but Pope has set such a pattern of ease in that way, that few composers now are faulty in the business of rhiming.

When I have the pleasure of meeting you, we will talk over these and many other matters too copious for an Epistle. I had like to have forgotten to add that Jack copied your verses and sent them to your Uncle John, who commended them much: you will be pleased to be commended by one that is the best performer and the best critic in that way that I know. With respects to your father and mother and all the family, I remain

Y^r affect. Uncle,

GIL. WHITE.

Nanny White mends apace. She is still at Newton.

On November 17th, 1774, Mulso writes:—

"Why do you think of going with your Brother into Lancashire in the winter? Will not the Hampshire cold suffice? It was odd enough, that on the very morning that I received yours, in which you complain of the snow, I had been revising a letter from you, in which you tell me that you had rode out every day to contemplate that beautiful meteor, which shows itself to advantage in your uneven country. I am sorry you change your note. No one bears time better outwardly; and yet I know by myself that time has made some advances upon you, for yesterday I was fifty-three. I have one pleasure however in this increase of years, it is a longer date of our friendship. As to not visiting you I declare solemnly that for one cause or another, it has not been in my power. . . . I hope your brother will succeed in his purpose of settling his son to his satisfaction. The provision for children is an arduous duty: You have escaped it. . . . Health attend you wherever you go. The love of all here attend you, and our best wishes for nephew John."

To Mrs. Barker. London, Novr. 26, 1774.

Dear Sister,—I have been indebted to you for some time in the letter way: but as I have lately written to Sam, I was in hopes that a letter to one of the family would express my regard for the whole, and excuse my other obligations for a time. My business in town is to meet my brother John, and to bring up Jack, who is grown so tall and large that it is full time he was settled in the world. A person who calls himself a mercer has been much recommended as a master; he is said to be a good domestic, family man, and in a thriving way: but when my brother came to talk of price, he demands a much larger deposit than was expected; so the treaty is at a stand for the present. My brother John desires me to tell you that he once thought of returning with his son by Lyndon; but as he has spent already in town a good deal of the time that he had allotted for absence, and nothing is concluded in his business, he now finds that he must defer that satisfaction to a farther day. Originally I intended to have met brother J. in town, and to have accompanyed him to Blackburn, and so to have spent the winter between that place and Lyndon: but just as I thought I had at last procured help for my church, my assistant was called into Devon, to return he knows not when:—"ibi omnis effusus labor." Sam will tell you the meaning of the Latin. Molly White thrives well at Selborne, and grows tall, fair and handsome, and is a fine girl. Nanny Woods also is very stout and hardy, and is a nutbrown maid. Poor Nanny White, who came to Newton in so deplorable a condition, has for these last fine weeks mended in a most marvelous manner; so that her friends about her have good hopes; and if she has no relapse will be again soon in a comfortable state: though London, I fear, will be no ways fit for her for some time. Berriman lies still in the same sad deplorable way, helpless and hopeless!

Lately I had a letter from Mrs. Snooke; in the former part whereof she says that she had not been so well for a long time: but in the postscript, written the next day, she adds that the gout came on in the night. The present cold weather, I fear, will pinch her. Winter comes on with hasty strides this year: and I begin to fear we shall have a severe one. Tell my niece Betty that I don't love snow now near so well as when I was of her age: I then thought it a very amusing, pleasing meteor.

All friends here are well except Mr. H. Woods, who is in a poor, dejected way: he says business will certainly kill him if pursued; that if he could be quiet he should be well; but that he is so embarrassed that he cannot possibly leave off. Jack is five feet 8 inches and threequarters high without shoes, and proportionably large. Pray tell Sam that I shall hope to hear from him in prose or in verse, in Latin, or English as he likes best. The insinuation that Mrs. Chapone is a papist is a foolish slander thrown out by somebody that envies her literary reputation: I have been assured since that she is an Italian stage-dancer. Brother Harry has got his roof on his new building and makes use of his new kitchen. By February he probably will have an increase of family; not twins, I hope, though our sister looks very big.

I am, with all due respect, to all friends,

Your affectionate brother,

GIL. WHITE.

To the Rev. John White.

Selborne, Jan. 5, 1775.

Dear Brother,—When I had read your letter through and found that no real damage ensued, I could not forbear smiling at your accident, as I figured to myself the vicar of Blackburn, the man who for eminence sake is called *the vicar* riding on a post-horse with fardles buckled round his waste. Jack with his long legs would straddle along with ease; but if I had been there, my short stumps would have followed through the snow "non passibus aequis."

When you have settled with Mr. Morgan I should be glad to hear how you came off; and Mr. Pinnock, who recommended that gentleman so far interests himself in the event, as to wish to hear whether it takes place.

Gossomer has from old times attracted the attention of the curious. Chaucer mentions it among other phænomena of nature not well understood, or to be accounted for; such as thunder, &c.

The *Tabanus bovinus,* I verily think, has no sting in the tail, or blood-sucking rostrum, but a musca-like proboscis. I have seen it suck the galled parts of Sir Simeon Stuart's working oxen,* without giving them any pain or offence; it abounds most in moist places, and sultry weather. The *Œstrus curvicauda* never lays its nits, I know, but in the warmer hours of the day; for my horses, which are in stable all day, and out at nights are never covered with those eggs at home.

Pray examine those little dancing *Dipteræ, Tipulæ* I suppose they are, that sport the winter through in fog, gentle rain, and even frost and snow when the sun shines. Some *Phalænæ* appear the winter through in mild weather, though the papilios are soon laid up.

I have heard twice from Mr. Pennant within these three weeks; he is very pressing and desirous that your correspondence should be renewed; and hopes, if any resentments ever subsisted, that they may all be forgotten. Indeed as he seems so earnest, and is a writer by whom brother Ben will be such a gainer, I could wish that you would make advances so far as to continue to write to him

* How little Selborne has changed since this was written may be inferred from the fact that quite recently (in 1896) oxen were seen at plough there by the present writer, who was also astonished to see wheat cut by the sickle in an adjoining parish.

on the general subject of Natural History. I have just read his 'Tours';* they gave me pleasure and information. He is composing a map of Scotland suited to the occasion, which was much wanting. Mr. P. did not go to the Orcades himself, nor Shetland; but sent a Scotch clergyman,† a missionary, he calls him, to both those places; and intends to avail himself of his journals. Mr. P. was last summer at the Isle of Man, and intends to give an account of that place also. He has now taken great pains to investigate Great Britain and its Islands, and will be well qualifyed to put the last hand to the 'British Zoology' in a quarto edition.‡

You are by this time, I presume, settled quietly down to your Fauna; and will be able to make dispatch in your larger articles, now you have finished the small ware of your insects. Frequent inquiries are made about your work; and the Nat[uralist] world expects good things at your hands. N. White is still at Newton, and recovers marvellously. J. Neal is in a most dangerous way; and Berriman lies still in a lamentable state. The day after you got home I had a cold but most pleasant drive into Hants; there was just snow enough on the ground to make the carriage wheels run easy. We have now summer-like weather, and this day the insects swarmed under sunny hedges; the business of generation seems, even now, to go on among them. Pray provide a good many cuts for your work; plates at present recommend a work. Mr. and Mrs. Bassat, who are here, join in respects. Pray write soon. My respects to my sister, who must be happy to see her son so tall and healthy.

Yrs &c. GIL. WHITE.

* The third and last volume of Pennant's 'Tour in Scotland' was published in 1775.

† The Rev. George Low.

‡ Such an edition was, in fact, published by Benjamin White in the following year (1776), as well as one in octavo, and it was of this that Gilbert White corrected the proof-sheets.—A. N.

On January 31st, 1775, J. Mulso writes:—

"Are you not now glad that you did not travel or anti-migrate to Lancashire? What a season would you have had to interrupt every natural enquiry? . . . I have looked into a vol. or two of 'Philosophical Transactions' (our library books), and have found honourable mention of you from Mr. Barrington. Have your further observations been *redde* (as they print the word) and well accepted by the Society? I should think they must, for you are an accurate man. But I know that it is a rule with them not to commend or discommend, but to give as it is given to them, and leave it to the use and judgment of the Public."

To the Rev. John White. Selborne, Feb. 1, 1775.

Dear Brother,—I have been unusually dilatory in my answer to your last letter; and the reason was, though I was much vexed and disappointed at your rebuff, which came so unexpectedly, yet I did not know how to come to your assistance. Mr. Pinnock at the same time mentioned a gentleman of the law: but that is a profession that you do not seem to affect. I remember a person in Mr. Freeland's shop* (one of the book-keepers) mentioned a House in Manchester that wanted a young man; but as my nephew was at that time sped, as I thought, I paid but little attention to what was said. I will continue to make what enquiry I can; but it will be best, no doubt, to urge all your friends in London. The task of providing a proper place for a young man is doubtless very difficult; since Edmund White remains still in his father's house unprovided for.

Brother Ben. says you must have as many plates as possible in your Fauna; for it is the fashion now "to look in picture-books." Insects being the most laborious, and

* In Chichester.

probably, to most readers, the most uninteresting part of your work, I am glad you have run through them. The birds and quadrupeds will pass off smoothly. I am sorry you will not work up your tour to Mogador into a pretty chapter; it is the fashion now to publish tours; besides some account of the person, manners, mode of life, of that monarch who at present sets all the naval powers at defiance, would take with many readers. The practice of gratifying such barbarians with elegant presents, and the Moors turning Dollond's perspectives into walking-sticks, would furnish matter of agreeable reflection. Dr. Johnson has just published his Journey thro' the western isles; I have read it, and you should read it. It is quite a sentimental Journey, divested of all natural history and antiquities; but full of good sense, and new and peculiar reflections. It does not at all interfere with Mr. Pennant's book. You are much to be commended for your intention of taking all your duty on yourself. Mrs. White is just come to Newton, and intends taking Nanny in a few days to lodgings near town. Poor young woman, she is marvellously recovered; has been on horseback every dry day the winter through. John Neal and dame Knight are dead; Berriman lies in the same sad deplorable way. Mrs. Snooke writes me word that she has been better than usual this mild winter. For some days past we have had great rains, and blustering weather: this morning the weather is very wet and stormy; the thermometer at 50°, the barom[r] at 28·7. Every sunny day insects abound; and in warm lanes, and under hedges the air swarms with them. Within doors woodlice, spiders and *Lepismæ* are in motion, and many *Muscæ* in the stable; and earthworms come forth every mild evening. So that in mild winters insects are not so much layed up as is imagined. Some *Phalænæ* fly also all the winter. Mr. Yalden has a bad fit of the gout; this is the second attack; the first was almost two years ago. Harry's wife by this time is probably in the straw. On Jan. 20th many rooks were caught, it is said, by a man near Hackwood-park. Their wings, as he affirms, were frozen together by a wet sleet then falling. Pray write soon. Respects as due.

Y^rs aff.,

GIL. WHITE.

Mrs. Isaac writes me word that her aunt Weston is dead intestate; and that her share will be upwards of £5,000.

The following entry was made at this time in the *Naturalist's Journal*—

"Jan. 20. Mr. Hool's man says that he caught this day in a lane near Hackwood-park many rooks, which, attempting to fly, fell from the trees with their wings frozen together by the sleet that froze as it fell. There were, he affirms, many dozens so disabled."

Fyfield, March 9, 1775.

To the Rev. John White.

Dear Brother,—As you have long experienced that I am not usually a tardy and negligent correspondent, you will, I suppose, conclude that something has happened to prevent my writing sooner, as really has been the case. I have had an heat and stiffness in my eyes from over much reading, that made writing very irksome for some time; they are now pretty well recovered again.

After your disappointment in town I was glad to hear by your last that you had a prospect of disposing of your son at Manchester; but now I understand that farther difficulties arise. The Scopoli from Mr. Pennant that you mention is at Selborne; and I will send it, if you desire it; but it affords no information.

As you rather complain of some reserve on Benjamin's side respecting your work, suppose you write to him, and ask him how much he will give you *downright* clear of the plates and printing for your copy; and then you will know your certain gain, and will run no risk. Anything in the naturalist way now sells well. Or if he chuses to go shares in profit or loss, enquire of him what proportion he should think would pay him for conducting the sale and publication. Booksellers have certainly a power of pushing books into the world; and it must be a work of great merit to obtain and make its way *invitis bibliopolis.* You mention also a want of books: might you not also apply to Benjamin to know on what terms he would furnish you with the *use* of books proper for your purpose 'til you had completed your Fauna? It is highly proper, it seems, to have a good many cuts. Mr. Curtis will superintend your engravings.

Mrs. Chapone sold her two first vols. for £50. Now she has made up a third from essays, poems, adventures, &c. and sold that to another for £250; so that it is expected the man will lose considerably by the purchase.

Many thanks for the copies of your 'Gib. Letters,' which are very entertaining. You have the advantage of me now, since you have taken away my amanuensis. I am disturbed that Mr. Shaw takes no manner of notice of the *Hirundines*; nor how far the *melba* and *hyberna* extend, as might have been expected from his opportunities at Cadiz and elsewhere. Pray let Capt. Shaw know, that if he comes to Alton I should be glad to see him. The spirit for natural history that you left behind you is by no means evaporated; neither is your mantle worn out.

Lever has opened his museum at half a guinea per head. Harry has got a fine roomy kitchen indeed, and will have a fine parlor over. This addition shuts all his buildings finely together; and nothing is to be regretted but the expense. Sister Harry has got another fine boy, whose name is

Edward. Nanny White is in a poor languishing way, still at lodgings near Vauxhall: Edmund White is gone on trial to Mr. John Hounsom, linendraper in Fleet Street. The father is to advance with him a fee of £250; and the master makes a merit of taking so little, and says that from a stranger he should have demanded £300.

When opportunity serves, pray read Dr. Johnson's 'Journey through Scotland,' and Dr. Burney's 'Tour through Europe to make enquiries into the present state of musick.' Thanks for your information about cotton-cups.

Should you not produce in your work a short comparative table of weather at Gibraltar, Selborne, and N. America? Kalm will furnish you with the barom., thermom., &c. of America. I herewith send you my best account of the cobweb shower of 1741. What is said of spiders shooting webs, and flying &c. in Ray's Letters is so much, that it cannot be transcribed. You should consult Ray's Letters.

When first I came I fully intended to have sent you my account of the cobweb shower; but this house is so full that I have no opportunity of being long enough alone to think accurately on any subject; so I must defer that part till I write again. We have continual wet weather; and farmers are sadly hindered in their spring crop: stormy and wet this day, March 11.

When Hesiod says that the chirping note of the *Cicada* comes from *under its wings*, he expresses himself thus—

> "ἤχετα τέττιξ
> Δενδρέω ἐφεζόμενος λιγύρην καταχεύετ' ἀοιδὴν
> Πυκνόν ὑπό πτερύγων"—(*v.* 584).

Is there not a Frenchman who claims this discovery?

Sure insects have been more abroad this winter than usual: and lately, in our little interval of fine weather, many species of *Muscæ* came forth. *Chrysomela Gottingensis* begins to come forth.

Brother Thomas, Molly White, and myself came down to this place on Tuesday last; on Wednesday next Harry's boy is to be baptised; and on Friday we are to return to Selborne.

I have just dug away forty loads of earth from the end of my kitchen, and have now set my house above ground in all parts.

Mr. Halliday* behaves very well, and improves so much, that his friends are well pleased with the pains that have been taken with him. His parts, though somewhat backward, and slow, promise to be solid.

Building is very infectious and catching; I am so pleased with Harry's new parlour, that I want to go home and build one.

A certain plea of license against the incumbent's taking all the duty in person can avail him nothing. Every man may, if he chooses, do his own business himself, certainly.

A flock of Spoon-bills was seen last winter near Yarmouth in Norfolk: one was shot and sent to Curtis,† who showed it to brother Thomas. This is a rare bird indeed in England, though common in Holland, and must have migrated across the German Ocean, no narrow frith, in spite of all that Mr. Barrington can say to the contrary. That gent. is got into some fracas with the R. S.: so that, I suspect, no more of my *Hirundines* will be "redde."

I will send you in my next what Chaucer says about gossamer; it is wonderful that so remarkable and prognostic a phenomenon should escape Thomson, the naturalist poet.

As America is at present the subject of conversation, it may be matter of some amusement to you to send you a

* Mr. Francis Alexander Halliday, afterwards Captain, R.N., married Ann, daughter of his tutor, the Rev. Henry White.

† William Curtis, the well-known botanist and entomologist. He was born at Alton, and probably was well known to Gilbert White.—A. N.

quotation from the *Medea* of Seneca, prophetic of the discovery of that vast continent.

> "veneant annis
> Secula seris, quibus *oceanus*
> *Pateat, tellus, Tiphysque novos*
> *Detegat* orbes; nec sit terra ultima Thule."

N.B. Tiphys was pilot to the Argonautic expedition; and a type of Columbus.

All friends join in respects. Yours affect.,

GIL. WHITE.

Sure your Fauna should sell outright for £100 clear of all deductions. Mr. Pennant gets that sum for his new edition of 'British Zoology'; and your work will contain much more new, original information. I want to see you the first of Faunists. With regard to anecdote and real natural history the less you borrow from books the better; you have a large fund of your own. Benjn will get very largely by Mr. P.'s Scotch tour.

Under date March 7th, Fyfield, the *Naturalist's Journal* records—

"Bro^{r} Harry's strong beer, which was brewed last Easter-monday with the *hordeum nudum* is now tapped, and incomparably good!"

To Samuel Barker. Selborne, March 30, 1775.

Dear Sam,—As I took no copy of my last hasty letter on poetry, I am not very certain how far I went on that subject, and what I omitted: however I think I said nothing concerning the *power* that masterly writers possess of adapting their *numbers* to their *subject*, or rendering the sound "an echo to the sense." Homer and Virgil, no doubt, enjoyed this faculty in great perfection; and have shewed wonderful instances of it: but then you must remember that fanciful commentators have over-refined on this power, and have found numberless beauties of this kind, which the authors neither perceived or intended.

The English language is very capable of being conducted to this perfection: and Pope in particular in his translation of the Iliad has frequently imitated the original most happily in this way. That gentleman in his Essay on criticism, which he published, as I remember, at sixteen years of age, has given several instances of this sort of power: as,

"And the *smooth* stream in *smoother* numbers flows," &c., &c.

But the finest instance that I remember in our own language, for several lines together, is in old John Dryden's translation of a simile in Virgil which, though I have not seen for these twenty years, I shall never forget on account of its singular elegance.

"As when a dove her rocky hold forsakes,
Rous'd in a fright her *sounding* wings she *shakes*,
The *cavern rings* with *clattering: out* she flies,
And leaves her callow care, and cleaves the skies:
At *first* she *flutters*: but at length she *springs*
To *smoother* flight, and *shoots* upon her wings."

". . . mox *aere lapsa quieto*,
Radit iter liquidum, celeres neque commovet alas."

In short, John Dryden is, to me, much the greatest master of numbers of any of our English bards: but then, contrary to most men, he never arrived at perfection 'til he was very old.

Rhime is in itself barbarous and Gothic, and unknown to the ancients, who would have despised such a jingle; but then it must be remembered that modern languages, being destitute of the beauties derived from termination and inflection, require some substitute. Besides some of our best poets have conducted rhime with such address, that it seems to fall in of its own accord without their seeking: and if rhimes are shackles, yet these people move so gracefully in them, that we would not wish to see them divested of them.

Blank verse is, no doubt, when well conducted, full of dignity; but then perfection in that way is so rare, that we never had but two or three poems that were worth reading. A desire of raising the diction above prose pushes men into fustian and bombast. Even the great Milton, the father of blank verse, is not always free from this vice; but ransacks the whole circle of sciences for a set of hard words and rumbling terms that make his readers stare.

As to Thomson (not Tompson) his Seasons are sweet poems, full of just description and fine moral reflections. But then this Scotch bard, through a desire of elevating his language above prose, falls also into fustian sometimes; and though he thinks much like a poet, is often faulty in his diction.

The Cyder, of John Philips, a didactic and Georgic poem in blank verse, is worthy your attention. This man dyed young; but had he survived 'til he had acquired a little more ease, and 'til time had somewhat mellowed his muse, he had been an excellent poet.

Somerville, quite in advanced life, wrote his 'Chace,' a poem full of warmth and spirit, and all the enthusiasm of a young sportsman.

Thus have I given you my crude sentiments in a hasty way on the subject of English poetry. If my remarks afford any pleasure or information, my intention will be fully answered. Venus, Jupiter, Mars, and Saturn appear now every clear night as it were in a line; but how and when am I to find Mercury? Had it not been for your Father, who showed him to me at Lyndon in April, 1760, for near a fortnight together, I should never have seen him at all.

Yours affectionately,

GIL. WHITE.

To Mrs. Barker.

[On the same sheet.]

Dear Sister,—I could have much wished to have spent part of last winter with you; but just as I thought I had got a gentleman to supply my church, he was called suddenly into Devon. Harry has got a large family indeed; brother Thomas and I were lately at the Xtening of his last boy, whose name is Edward. Our brother has lately much enlarged his house, which could no longer contain his numerous family: a new kitchen, and a new parlour over that, and garrets over that, all very large and roomy, make the house now very commodious; and nothing is to be regretted but the expence. As building is catching I also talk of some addition to my house next summer; but I much suspect my resolution for setting about it. Brother John was disappointed in placing his son in London, and now thinks of placing him with a linendraper at Manchester; a scheme, I think, much for the better in all respects. For in London they ask most enormous fees; and Bro. Ben has just given £250 with Edmund* to Mr. Hounsom in Fleet Street. Alice Boxal, who removed after her husband's death to her daughter's, is lately dead; as is also John Neal. Poor Berriman lies in the same deplorable way still! Nanny Woods continues stout and well, and is a fine brown maid: her hair is remarkably fine. Mr. Woods never comes to see her. My brother Thomas thinks he will now certainly decline business. Mrs. Snooke, as often as I have heard from her, has acknowledged herself better this winter than usual. Molly White is very well; and is stout and large of her age, and a giant to Mrs. Etty. Your kind present

* He subsequently quitted trade for Oxford and the Church, settling at Newton Valence vicarage as his uncle's neighbour.

to your native place* I have disposed of in part; such gratuities in these hard times are very acceptable. Mrs. Isaac writes me word that her Aunt Weston dyed intestate, and that by standing in her mother's shoes she shall come in for about £5,000. Altogether her children will be finely provided for. I am concerned to hear that Mrs. K. Isaac has such poor health.

With respects to all the family, I conclude,

Dear sister, your affectionate

and obliged brother,

GIL. WHITE.

Mr. and Mrs. Etty and Niece join in respects. Friends are well at Newton. Brother John is in pretty good forwardness with his 'Fauna Calpensis,' or Natural History of Gibraltar.

I forgot to tell my nephew in the proper place that Dryden's ode on St. Cecilia is nothing else for an hundred lines together, but beautiful numbers finely adapted to the sense. He will, I hope, write soon.

Fierce frost at present with snow; woe to the wallfruit!

On May 14th, 1775, Mulso writes:—

I am diverted with your effort to ally me to Sir Kenelm Digby. Could you prove me next heir to his genius or estate, it might do something. But as to the precarious relation to the Mulsos of Gothurst in Bucks—tho' it wears a face—*vix ea nostra voco* . . . As for your itch of building, nothing cures it but experience. You would have great pleasure and pride in an essay or two, but upon a *repetatur idem* you would shrug your shoulders. But, nevertheless, try—for the Devil of Taste will haunt you in your sleep—"*Aude aliquid—Carcere dignum.*" It will go off in one

* Mrs. Barker annually sent her brother one guinea for the poor of Selborne.

fit of Pride of your Performance. I am going to see whether the Hambledonian has furnished me with another flower that will puzzle you. Come and see. I believe I have a great deal to say to you. Old friends cannot meet without confab. I constantly feel a disposition to welcome you, being ever, my dear Gil., Yours affctly.,

J. MULSO.

After visiting his friend at Selborne, who took the opportunity to read some of his compositions in natural history, he writes again on July 8th, 1775 :—

"I do not know that ever I made a visit more to my satisfaction than this that I have just finished to you. . . . You have a double felicity in your manner of entertainment; you can gratify your visitors both with beautiful originals, and high descriptions; representations studiously copied from nature, and finished with a masterly hand. As you intend your works for the public, I would not say so much in a strain of flattery; for though I would not tell an author how much I liked his productions, yet I might slubber them over with a hasty careless compliment, or lose them in silence. . . . You have happily grounded ethics on a stable and beautiful basis, the works of God; and your figures, formed from naked and genuine beauty, beat every finical composition that would fascinate the judgment by adventitious ornament. This is my real opinion of your Work. But mem. I do not mean by the close of the last sentence a slur on your intention of employing the art of Mr. Grimm, or any other more accomplished designer. I wish he may add to the pleasure of the world, as much as he will gratify my partiality, if he can convey . . .* your truly delectable scenes."

* Letter imperfect.

A little later her nephew visited Mrs. Snooke at Ringmer, whence he wrote—

To the Rev. John White. Ringmer, Aug. 12, 1775.

Dear Brother,—As you had once experienced a disappointment with regard to the disposal of your son; I was much pleased to find by your last that he was now finally settled under a master of so much reputation;* and I doubt not but that Jack, who has abilities, will hereafter make a good figure in life. Moreover, as you have but one child, I think it is much more comfortable for you and my sister to have him fixed near at hand, than in London, where you would be able to see him so seldom, and to know how he went on. As to Jack's "venturing to draw blood from his majesties subjects," I do not so much wonder: I rather admire at the courage of the patients who permit him: however every young man must have a beginning.

Mr. Aikin† I have heard of, and seen extracts from some of his writings; when he offers to correct the press, how are you to gratify him? It will be very clever to have such a corrector.

Mr. Grimm, the Swiss, is still in Derbyshire; and is to continue there and in Staffordshire 'til the end of the month. I have made all the enquiry I can concerning this artist, as it much behoves me to do. Mr. Tho. Mulso, and Brother Thomas, and Benjamin, and Mr. Lort have been to his lodgings to see his performances. They all agree that he is

* A surgeon, probably at Manchester.

† John Aikin, M.D. (1747–1822), was at this time practising as a surgeon at Warrington. He subsequently went to London and devoted himself to literary undertakings, his reputation chiefly resting on his endeavour to popularise scientific inquiries. He is mentioned by Gilbert White in the last sentence of 'The Natural History of Selborne' as having written 'The Natural History of the Year.' In 1795 he wrote a preface to and edited 'A Naturalist's Calendar,' etc., from Gilbert White's unpublished papers. It is unlikely that he was the surgeon to whom John White, jun., was apprenticed, since he and his family were strong Dissenters.

a man of genius; but the two former say that he does hardly seem to stick enough to nature; and that his trees are grotesque and strange. Brother Benjamin seems to approve of him. They all allow that he excels in grounds, water, and buildings. Friend Curtis recommends a Mr. Mullins, a worker in oil-colours. Grimm, it seems, has a way of staining his scapes with light water-colors, and seems disposed much in scapes for light sketchings; now I want *strong lights and shades* and good trees and foliage.

How apt is each person to think better of others' circumstances than his own? you say my "letters are all written": but I say that you are in much more forwardness, for your Fauna is just finished: while my letters, many of them, want transcribing; and much writing gives me a pain in my breast, and I can procure no amanuensis; my journal is but just begun; and the antiquities of Selborne are not entered upon at all. Friends in Oxon., I hope, are searching for me amongst Dodsworth's collection of papers in the Bodleian library, 60 vol. folio; but the papers that I want to see most are immured in the Archives of Magd. Coll. "de mercatu, et feria, registro," &c., &c.

Suspecting from its habit and shape that the fern-owl might resemble the cuckow in its internal construction, I procured two; and found my suspicions not ill grounded. For upon dissection the crop or craw* lay *behind* the *sternum*, immediately *on* the bowels. It was bulky and stuffed hard with large phalænæ of several sorts. Now as it appears that this bird, which undoubtedly sits itself, is formed exactly as cuckows are: we may reasonably conclude that Mr. Herissant's conjecture, that cuckows are incapable of incubation from the disposition of their intestines, becomes groundless; and we are still to seek for the cause of that

* Gilbert White uses the words "crop or craw" in a sense quite different from that now applied to them. Neither the cuckoo nor fern-owl has a "crop," and what he so called was simply the stomach.—A. N.

strange peculiarity. Sam Barker and I caught a female viper the other day. When we came to cut it up fifteen young as large as the biggest earthworms came issuing into the world gaping, and setting themselves up, and menacing in a most extraordinary manner. I greatly suspect the fact about the dam's admitting them down her throat in times of danger: for unless we confounded the gullet with the windpipe, the passage we saw was not half big enough to allow of room for the young: therefore if they ever retreat into their mother's body at all, it should rather be at the other end, thro' the anus, which is wide, into the abdomen which is very long and capacious. The vitals and viscera lie in a very small space in the middle, and the gullet and colon of course are very long.

At my arrival I found here Mrs. Barker, Sally, Molly and Sam; Mr. B. came since. The young people are clever and intelligent. Mrs. Snooke is very well, and a marvellous woman at 81. We have sweet harvest weather, and a noble crop of wheat, and fine hops at Selborne; though they totally fail round Canterbury. Here are no cross-bills this summer in Mrs. Snooke's pines; some years they abound. Sam Barker wishes to be a naturalist. Next Thursday I must set out for Hants; the Lyndon people [*i.e.* the Barkers] will follow in a week or ten days. I rejoice to hear that you are so well satisfyed with your new curate, and hope he will not find the duty too much for him. Jack's box was sent up to Edmund. All friends join in respects to you and my sister. Pray write soon. Yours affectionately,

GIL. WHITE.

Soon after his arrival at home again the *Naturalist's Journal* records one of those little experimental investigations which were constantly made by the man who was ever observing and ever noting all natural facts that came within his purview.

"Sept. 30. My *Arundo donax*, which I received from Gibraltar, is grown this year 8 or 9 feet high. I therefore opened the head of one stalk, to see what approaches it had made towards blowing after so hot a summer. When it was cut open we found a long series of leaves enfolded one within the other to a most minute degree, but not the least rudiments of fructification: so that the plant must have extended itself many feet before it could have attained to its full stature, and must have many more weeks of hot weather, before it could have brought any seeds to maturity."

To the Rev. John White. Selborne, Octr. 4, 1775.

Dear Brother,—From the hurry arising from a full house while the Lyndon family were with me; and by means of Mr. Thomas Mulso, who came as soon as they were gone to Fyfield, I find that your letter has lain unanswered for three weeks. It is proper therefore to sit down now I am alone, and answer your last before my friends return from Fyfield. Mr. Barker sets out as this morning for Northamptonshire, and takes his leave of Hants at my brother Harry's house; but the ladies and Sam return hither on Friday, and Harry accompanies them, and stays with me a few days. How long my Sister, &c., are to stay I cannot yet say.

Your Fauna, to which I think myself at least a foster-father, is become, I hear with pleasure, a fine thriving child. I could be glad to examine its features, and to dandle it, and remark how it shoots up towards its ἡλικια; but the old difficulty of my Church stands still in my way, and is like to prove as great a *remora* as usual: I am making enquiries concerning some assistance, but can hear of nothing yet to my satisfaction. No Grimm has yet appeared: the reason is because he has been detained so long in Nottinghamshire. . . .*

* A considerable part of this letter is missing. It continued with an account, evidently copied from the foregoing entry, of the *Arundo donax*.

Our people here, you know, call coppice-wood, or hedge-wood, *rice*, or *rise*. Now brother Thomas has found that this word is pure Saxon: for *hris* signifies *frondes*; thus has he vindicated this provincial word from contempt. I am lowering my bank in my garden, and throwing its border on an hanging slope: last winter I sunk my walks so much, that this alteration became necessary. Where is Wollet the draughtsman to be found? Thomas Mulso, who draws sweetly, has taken hints concerning Hawkley slip, to be finished in town. Brother Ben. has just purchased two freehold houses in S. Lambeth, one of which is to be used as his country house, into which he is to enter as soon as possible. He and Ben. [his son] are just gone from us: my Sister Ben. and Jane and Nanny are still at Newton; the latter is most marvellously recovered, and will now, I trust, do very well. Poor little Nanny Woods has been ill, and has lost her colour. Brother Harry has got an other young man, a 50 pounder. He has now a fine income; and will soon, I hope, begin to lay by some money. Does your migrating clergyman visit you again this winter? Ring ouzels came to us in September. Your snuff-pincers extinguish my candles in a very neat manner.

With respects to my sister, I remain

Your loving Brother,

GIL. WHITE.

The following letter shows the interest in his brother's book that the writer took and the assistance which he was able to render in some points :—

From Thomas White. London, Nov. 9, 1775.

Dear Brother,—I shall be glad to see you in town, but know not what to say concerning the disorder that is very

A REMINISCENCE OF SELBORNE IN 1751

From a Drawing (probably made in London after a visit to Selborne) by Thomas Mulso

[To face p. 292, Vol. I.

general here. I believe most people that have it felt ill some time ago; but I am not conversant enough amongst sick people to say positively there is no fear of your taking it now. Molly and the whole family have had colds, coughs, &c., but are now nearly well; as to myself, I have escaped, like John Wood's old horses, by old age and other infirmities. Thank you for the elegant quotation from Middleton.* Is not the ridicule some of our wise governors would have thrown on America applicable to Cicero's on Britain? and may not America be to England ere long what England is now to Rome? I cannot allow that the Romans acquired their riches by virtuous industry; the infamous oppression these people exercised over mankind has been handled too tenderly.

Hlinc is pure Saxon, a bank cast up for boundary; hence our "linch" and "linchot" between fields. As you seem to allow me to frolick in conjecture (as Johnson says), I will examine the fields.

Molly goes to-morrow with Dr. Thomas to Cambridge; she has had no return of her complaint, and is to use the cold bath there. I want you to read Plot's treatise 'De origine Fontium,' in which he states what has been advanced on all hands by former writers and favorers, the assertion of subterranean connections with the sea, against Ray and others. I cannot help looking on these communications as imaginary, and am inclined to join with Ray, who asserts that rain and dew are sufficient to supply all springs. When you describe the perennity of the Selburn spring, it does not seem foreign to the purpose for you to sum up the evidence on both sides, remarking the peculiarity of upland ponds being supplied when those in the vallies fail, which I believe will prove a new observation. Certainly hills and mountains

* The Rev. Conyers Middleton, D.D. (1683–1750), author of a 'Life of Cicero.'

are condensers, and convert by their coldness the ascending vapours into water; but more of this when we meet.

I am, yours affy.,

THOS. WHITE.

To Samuel Barker.

Selborne, Novr. 15, 1775.

Dear Sam,—After some consideration I am in no manner of doubt but that "murmur *electricum*"* is an error of the press; and that it should be murmur *elasticum*. For what in the world has electricity to do with hop-poles? why, if it had, should the *wind* call it forth? Now as to an *elastic* murmur or a deep *humming* sound occasioned by the vibration of the *naked* poles when agitated by the wind, I have heard it 20 times in the months of March and April: and moreover when I came to question my servant Thomas, he readily recollected to have heard such a *rushing* in hop-gardens in the spring-months; and added, pertinently enough, that such a *murmur* might be observed every spring in gardens among kidney-bean sticks; as I perfectly well remember. Therefore read, meo periculo, *elasticum*, instead of *electricum*. The only thing that sticks with me is, that since this murmur may be so easily and naturally accounted for by *elastic vibration*, why should Linnæus express any wonder, or be in the least pother about so plain a matter? since it seems to me, that it is as obvious why a pole should *hum* when put in brisk motion, as why the strings of an Æolian-harp should, when brushed over by the wind, produce those delicate chimings and unisons; that is by vibration. It is most probable therefore that there are no hop-gardens

* *Vide* Linnæus's 'Species Plantarum,' second edition (1763), p. 1,457, "Murmur electricum quasi remotissimum tonitru vento exagitante Humuli palos quid?" From information kindly supplied by Mr. J. E. Harting, it appears that no correction has been made in Linnæus's own annotated copy, now in the library of the Linnean Society, Burlington House.

in Sweden; and that Linnæus never was witness himself to such a murmur; but takes his cue from some hasty and inaccurate correspondent.

I beg you would take two pieces of spunge of equal size, weight and softness, and hang them by strings over an upland pond *in foggy weather*, the one as *near* the *surface* as *possible*, the other several *feet above* the water: then I desire you would squeeze the spunges in a *morning*, and see which produces the most water. Now if the lower spunge should prove, from repeated trials, to be the moistest, I should hope the fact would in some measure *corroborate* my suspicions, that ponds, and pools do by *condensation* from the *coolness* of their surfaces assimilate to themselves fogs and vapours by contact and that is one reason why many very *little* upland ponds, though subject to a continual waste by cattle, &c.; yet never fail in the severest droughts, while larger pools in *bottoms* frequently become dry. But as your father and you may most probably hit upon a better experiment, I desire you would try such as you think most to the purpose. Moreover, I desire that both of you would send me *every hint* in *Natural History* that occurs to your minds after your recent visit to these parts. My swallow monographies are printed off by the R. S. in Vol. 65, p. 258; but the corrector of the press has made sad work with my unfortunate letters; for in one place he makes me say that "swallows eat *grass*" and in an other uses *caves* instead of *eaves*: moreover he has transposed my letters so as to misplace them, though I numbered them most exactly, and by that means has made a jumble of dates; besides putting two *whens* in one sentence; and many more inaccuracies too numerous to mention! O fie! for so young a man as you to use glasses that magnify 200 times, when Linn. planned and perfected his whole sexual system *nudis oculis.* I wish you joy that Jupiter is restored to his liberty and dignity, for the Cornish man has seized on him, and appropriated him to himself as a new-discovered world. Mrs. Etty's inoculation is in a most prosperous state: only a few pustles to each party. Mr. Etty and I live here by ourselves, and having no wives to controll us, do as we please: only we are deterred from going to London by the influenza. Pray return my best thanks to my sister for her agreeable present; and to your father and sisters for their company and conversation at Selborne. I acknowledge myself much in your debt and shall endeavour to pay you in kind. Did you find *rushes* as much in use at Lyndon as Mrs. Rashleigh* has done at Penquite? Many gentlemen in Oxford had never heard of rushes; perhaps because they *were* gentlemen. Yours affect.

GIL. WHITE.

A few days previously, on November 1st, he had written what became Letter XXVI. to Barrington, on the use of rushes instead of candles.

* Formerly Miss Catharine Battie.

CHAPTER XIII.

EARLY in 1776, owing to the death of an annuitant, Thomas White came into full possession of the property settled on him by the will of Mr. Holt, who had died in 1745. Subsequently he quitted London for the then rural village of South Lambeth.

To Thomas White.

Selborne, Jan. 4, 1776.

Dear Brother,—As I have often heard Sir S. Stuart say, that if he had his timber to sell over again he could sell it for five or six hundred pounds more than he made of it; and as men seldom have much timber to sell a second time, you should, I think, retain Mr. Hounsom as your council, and make use of his superior judgment before you bargain. I hope you will find £2,000 worth of trees that are ripe on your estate; and that sum will help much towards your younger children's fortunes.

By all means, when you are more settled, begin laying in a fund of materials for the Natural History and Antiquities of this county. You are now at a time of life when judgment is mature, and when you have not lost that activity of body necessary for such pursuits. You must afford us good engravings to your work, and carry about an artist to the remarkable places. In many respects you will easily beat Plot: he is too credulous, some times trifling, some times superstitious; and at all times ready to make a needless display, and ostentation of erudition. Your knowledge of physic, chemistry, anatomy, and botany will greatly avail you. The sameness of soil in this county will prove to your disadvantage; while the variety of Staffordshire * is prodigious; Coal, lead, copper, salt, marble, alabaster, fuller's earth, potter's clay, pipe clay, iron, marl, &c.: while we in general have nought but chalk. But then you must get Benjamin to write abroad for the treatise 'De cretâ,' and make the most of it, as it is so little known. Bishop Tanner† will be of vast use for the religious houses. It is to be lamented that Plot was prevented by death from going on: for he improves vastly in his second history which greatly exceeds his Oxfordshire. We have, you know, an actual survey of Hants which you must get reduced so as to fold into a folio. You should study heraldry, and give the coats of arms of our nobility and gentry; 'til lately I was not aware how necessary that study is to an antiquarian: it is soon learnt, I think. There are in this county 253 parishes, most of which you should see. The Isle of Wight must also come into your plan.

Time has not permitted me yet to go through half Priestley's Electrical History, but in vol. 1, p. 86, I remark, that Dr Desaguliers proposed the following conjecture concerning the rise of vapours. "The air at the surface of water being electrical, particles of water, he thought, jumped to it; then becoming themselves electrical, they repelled both the air and one an other, and consequently ascended into the higher regions of the atmosphere." If

* Of which county, as well as Oxfordshire, Dr. Robert Plot wrote 'The Natural History.' It is no doubt from these books, which were well known to him, that Gilbert White took his title of 'The Natural History of' his native parish.

† Who wrote 'Notitia Monastica.'

Thomas Robinson pinx. Walker & Cockerell ph. sc.

Tho. White

this be *always* the case, what becomes of *our* supposition, which is, that by contact and condensation the water in vapour is drawn *from* the *air* to the *water;* and that thus upland ponds are mostly supplyed?

I never saw an electrometer. Our neighbourhood is all bad with colds; and among the rest myself also: some have eruptive fevers.

Y^rs affect.,

GIL. WHITE.

To the Rev. John White.

London, Jan. 30, 1776.

Dear Brother,—As you have enjoined me to speak my sentiments with respect to your work, you must not think me didactic and forward in the following pages.

It will be well to sweeten and diversify your tables of weather &c with an alternate page of zoological calendar, and interesting coincidences: for the generality of readers are apt to skip over whatever looks like figures. Your Journal will be pretty long.

An index perhaps has never entered your head: yet such a thing may be expected in so large a work.

You may no doubt, if you please, invert your system as well as Brown. You are not sworn to follow the arrangement of Linn.: by that means the subject certainly rises on the reader. The Swedes admire Brown notwithstanding. *Faunæ Calpensis primitiæ* will no doubt be more modest; yet might your *full* History well deserve to be called a Fauna. In strictness Linn.'s 'Fauna Suecica' is no more a *perfect* Fauna than your own, since some hundreds of animals have escaped his observation. Brother Ben. objects to a Latin title to an English book. Suppose you call it 'Fauna Calpensis or a Zoological History of Gibraltar &c.,' for Natural History it must not be called, since the plants are wanting. There is such a spirit gone forth against whatever is Linnæan, that I would not make the title page Linnæan.

Your Bookseller must be consulted a little in the title page and advertisements; as he knows best how to throw in little savoury and alluring circumstances to quicken the appetite of your buyer. By no means should you print, brother Thomas and I both think, 'til you have sold your copy: booksellers know how to subscribe off an impression to the trade, and to throw cold water on a work lying on the author's hand. We do by no means like your "*sequimur patrem*" &c. You should have mottos relative to each class. Ovid, perhaps, somewhere among his monsters will furnish for the *Vermes.* Pray correspond with Padre Floroz, since Linn. will no longer write. We can by no means see how you can be off from bringing up your work yourself: for no person will purchase what they have not seen: besides one hour's conversation will do more business than an hundred letters. Might not Benjamin print and publish for you on the usual terms? We wish to see your papers, and to correct here and there, not out of vanity and a meddling temper; but because little errors unavoidably befall and escape every Author. Be sure procure a good perspective *western* view of the harbor, town, and hill of Gibraltar to fold in as a frontispiece with references: it will contribute to explain many passages. You will have, I find, near a 1,000 pages, and 800 species.

As an Electrician you should see Priestley's History of electricity. He sets the whole in a pleasing light.

Just as you wrote last my neighbours told me there were troops from Gibraltar at Alton; so I sent Thomas over with a note and invitation to your friend Capt. Shaw. He brought word that Mr. S. had passed through Alton, and was quartered at Farnham. I then wrote by the post to Farnham renewing my invitation; and received a letter of thanks, and excuses that they expected daily to be called for to march towards London, where the King would review them. However, the deep snow came and stopped their march; so that when I came to Farnham I found the soldiers still there. I therefore called on Capt. Shaw for five minutes in my post-chaise at the Bush-inn-gate, and so ended the affair. He express'd his sense of my civility; and says he will write to you very soon. He does not expect to go to North America.

As soon as brother Thomas is at liberty he proposes laying in materials for a History of Hants; he is in possession of a fine fortune.

Brother Harry I find has given you an account of Mr. Holt's will; had no will appeared we might have received £1,000 each. It is possible that the bequest to the lying-in (lying he calls them) hospitals, of which there are seven, may become a lapsed legacy; because the executor, who had a discretionary power with respect to them, is quite blotted out. But the general notion is that Ld Chancellor will supply that defect. Should he not interpose we may possibly have yet £300 or £400 each.

I am glad John (for now he is very near six feet high I must no longer call him Jack) behaves so much to your satisfaction. He has lately written to me; and I have answered from hence offering him a book of a guinea value, but desiring him to consult you and his master.

We condole with you on the loss of your excellent friend the Governor.* Perhaps by permission you may dedicate to the Archbishop; and as the General is dead may be allowed to speak with more warmth of him, than you could have done to him. Shall I desire Mr. Lort to inquire whether such a dedication would be permitted, and well taken?

* The Hon. Edward Cornwallis, a twin-brother of the Archbishop of Canterbury, through whose instrumentality John White had obtained the living of Blackburn, of which the Archbishop of Canterbury was the patron.

Poor Brown* the artist! it is the fate of most ingenious foreigners; they have no manner of economy. Forster† will be soon in the same condition; he and his son dress like noblemen, and give £60 per ann. for an house! They have published 'New Genera of Antarctic Plants.' Benjamin has a share in this book: there is *Barringtonia*, a *Sheffieldia*, a *Skinneria*, &c., &c. Their great work or 'Voyage' is now under correction at Oxford.

Have your churchyards in the North any yew trees?

Pray send me Reaumur's whole account of the *Hippobosca hirundinis*. Pray write soon.

London is now Petersburg; it freezes under our beds with shutters closed and curtains drawn. Brother Ben's new house at S. Lambeth was last Sunday Archangel, with the thermometer at eleven, and everything ice and snow.

My love to my sister,

Y^rs affect.,

GIL. WHITE.

Look in Anacreon's 'TETTIΞ,' ode 43, and see if it affords any apt motto for insects in general.

I have been to Mr. Grimm, and am better pleased with his performances than I expected; and think I must send for him next summer. Brother Thomas talks of employing him some time hence.

The greater part of the following letter refers to a purely private matter—the distribution of the personal estate of Mr. Thomas Holt, great-uncle (in the half-blood) to Gilbert White. It is printed as showing the shrewdness and caution with which

* Peter Brown, a Dane, author of 'New Illustrations of Zoology,' published this year, and mentioned by Pennant ('Lit. Life,' p. 25).—A. N.

† J. R. Forster, who sailed with Cook on his first two voyages, being accompanied on the second by George Forster, his son.—A. N.

the latter invariably approached all business matters that came in his way :—

To Thomas Barker. Thames Street, Feb. 7, 1776.

Dear Sir,—Our snow, like yours, was not very great, but most marvelously drifted through the hedges, so as to fill all our lanes and cover the gates. I was forced to dig my way out of the village and to ride for two Sundays following to Faringdon attended by pioneers. As long as I stayed, the thermometer continued abroad at 20°. But on Jan. 28th (Thomas writes me word) it fell to 7°, on 29th to 6°, on the 30th it was at 10°, and on the 31st it descended half a degree below 0° !! a degree of cold beyond any instance that I have yet heard of! There was a rime. At S. Lambeth it was at 7°.

Mr. Holt's blotted will continues still in the same dubious state. You will please to observe that his *heir* at *law* (supposing the will invalid) will have nothing to do with his effects, which are all *personal;* but his next of kin, viz. ourselves and Mr. Tawke and Mrs. Tuck, and, as Mr. Butcher says, two sons of a Mr. Ch. Holt, who like ourselves are all descended from the same common ancestor, our great Grandfather Mr. Holt of, I think, Petersfield, who had *two* wives and *two* families: but with this difference, that we stand only in *half* blood to Mr. Holt of *Thorney*, and the other parties in *whole* blood. But this makes no difference with respect to *personals*; for *half blood* shares equally with *whole* in personals: of this I have known instances. Mr. Holt, after legacies of £700, bequeaths *all the rest* to the foundling and lying-in (lying he calls it) hospitals: but adds, that if there should at his death subsist more than *one* lying-in hospital, that his executor shall at his own discretion proportion their shares. Now it happens that there are *seven* lying-in hospitals, and the executor's name is so totally erazed as not to be legible at all. Thus you see if we have any chance *at all*, it must arise from the shares of the *lying-in* hospital in the nature of a lapsed legacy. Moreover, you are to understand that Mrs. Woods' children can have no claim, since they are removed one generation from the *nearest of kin*; and there is no court where the Judges will *prædetermine*; since if they did, Lawyers must lay aside their coaches and four. Mr. Tawke talks now of administring. As to myself I expect nothing, if we meddle, but *seven* chancery suits on our backs. I wish I may be so fortunate as to express myself intelligibly on this intricate subject; and I also heartily wish that the poor gentleman (who no doubt had full power over his own effects) had either made a will or no will. This matter probably may not be settled for many years.

Thanks for your information concerning Mercury. Pray throw out *all* sorts of Nat. hints. I have employed the keeper of Domesday book to transcribe all relating to Selborne; and am to pay 4[d] per line: besides I have applyed for a transcript of all relating to the priory in Magdalen College archives. Pray send me word what rivers were frozen in Italy; were they in Lombardy or in the more S. parts? Virgil in the Georgics mentions freezing rivers. Mr. Gibbon, a Hants gentleman, publishes next week his first volume in quarto of a history of the latter Roman empire: there will be four volumes in all. I conclude respectively,

Your affectionate Brother, Uncle, and Brother,

GIL. WHITE.

Respects to family. Mr. Gibbon begins with Trajan. Pray write soon.

Though Gibbon's father was resident so near Selborne as Petersfield, it does not appear that there was ever any intimacy between the historian and Gilbert White, or indeed that they ever met, though very possibly this may occasionally have happened at the houses in the neighbourhood.

To Mrs. Barker.

Thames Street, Feb. 7, 1776.

Dear Sister,—Mr. Etty and Charles and I came to town on January the 22nd and 23rd, and found the public roads better than we expected. Mr. Etty brought up his son, and has placed him at St. Paul's school, hoping that by means of good friends he may procure him some considerable exhibitions, that may help to support him in an university education.

N. Woods returned a long while ago to her nurse's house: she and all the numerous party had the small pox in the most favorable manner. As to Mr. Woods you must not wonder or resent because he does not write: for when his daughter had a fever more than a year ago, Mrs. Etty wrote to him *every post* for many weeks; yet he never returned one answer. However, when all was over he sent her a very handsome set of tea-china. And now during inoculation he never wrote once in the time or since: but is going to make a very handsome present of plate. Thus you see some men are of an *un-writing* constitution, and yet neither want gratitude nor generosity. Why Molly White does not write to you I cannot tell, unless it be because she is niece to H. Woods. She is still at Cambridge and pretty well.

Brother Thomas talks of leaving off, and is letting down his stock in trade by degrees. Brother Harry I found here; he was snowed in for a fortnight or more; and when he went away on Jan. 25th wrote word that it was with the utmost difficulty that he got along; and that if he had attempted a week sooner he would have been stopped by the way.

Nanny White, so far from being over-much affected by the severity of the season, bore it wonderfully well. I have been spending some days at my brother Ben's new house at S. Lambeth, which is very commodious, and very handsomely furnished. The rooms are rather small; but my brother has removed the partition between two of the parlors, and so has made one good sitting-room: he has besides another parlor below, and a drawing-room above stairs of 27 feet in length; but it is narrow. This room my sister has furnished in a splendid manner. In short they, who have eleven children, shame me who have none, and yet make a pother about building one room. Nanny recovers very fast by living in the country; and my sister looks much the better for being out of town. In short this house will probably lengthen all their days. Poor Berriman lies still in the same sad state! Farmer Parsons has been near death with a fever, but is better. Abram Loe is dead, and has left a Widow and five small children, who are supported by Emshot parish. Farmer Turner has left his ale-house and is retired to his new house. He is before this, I trust, marryed to his housekeeper Rose Rawkins; the Bridegroom is 71 and the bride 69! I used to say that female beauty does not last above a century; but now I begin to retract. Betty Stevens is dead, and Thomas's mother past all hopes. I did not send for Baptist Isaac at Xmass, but hope to see him at Whitsontide. Mrs. Snooke, I fear, must suffer."

To Samuel Barker.

[On the same sheet as above].

Dear Sam,—I thank you for your kind and intelligent letter; but you never told me whether the good people of Lyndon *burn rushes*, or not; nor what you think of my amendment of *murmur electricum* of Linn. Dr. Hales, in 'Vegetable Statics,' settles the point from experiment, that the *moister earth is*, the more *dew* it attracts; and that a surface of *water* attracts more moisture from the air than a surface of *moist earth.* I wish I could assist you in mosses: time and practice will render you more adroit; but some lessons would be better. Dr. Forster has already published a quarto vol. of Antarctic genera; new genera, with cuts. It is a splendid book in Latin, and dedicated to the king. Brother Ben. has a share. The Doctor has done honor to his friends; and has got a *Barringtonia*, a *Banksia*, a *Sheffieldia*, a *Skinneria*, &c. The *Barringtonia* is a noble flower, and is *polyandria monogynia.* The Latin, I think, is good. His nautical work, in English, is now under correction, as to stile and idiom, at Oxford. Pray be more explicit about "the influence of the W. and S. exposure on our trees." We that see them daily do not so much observe. Write very soon and direct to this house. Send me *all manner* of hints for Natural History. I have made a visit to Grimm. Brother John has finished his Fauna; the work will be large. Mr. Barrington wants me to join with him in a Natural History publication, but if I publish at all, I shall come forth by myself. Brother Thomas is laying up materials for an History of Hants; he will some day take an artist down.

The first part of the following letter relates to the Chancery suit resulting from Mr. Holt's will, which seemed likely to rival the celebrated one of "Jarndyce *v.* Jarndyce" in length. A letter of the same date, and on the same subject, addressed to Mr. Thos. Barker, ends:

"Perhaps by and by the Ld Chancellor may be of our *first* councils opinion, and may decree away everything from us. So uncertain are all human concerns!"

To the Rev. John White. Thames street, Feb. 27, 1776.

Dear Brother,—Since I came up I wrote you a very long letter, and was in hopes of an answer.

Our interest in Mr. Holt's effects wears a better face than when I wrote. For now a second council (Mr. Madocks) says point blank that the legacy intended for the *lying-in hospitals* is *lapsed* to all *intents* and *purposes* by the erasure of the executor who had a discretionary power of proportioning their shares respectively; there are *seven* lying-in hospitals. Moreover the commons, before whom this matter has been pleaded, are of the same opinion with Mr. Madocks. Mr. M. farther hints a suspicion of the invalidity of the whole will; but on that event we must not presume; most probably the foundling Hospital will be secure of about £4,000. Mr. Tawke has acted in an underhand manner by endeavouring to *steal* administration by himself; but the commons and foundling both seem to declare that he shall *never* administer, but that the Whites shall. Supposing by and by that chancery does not take all away by declaring the claim of the lying-in Hospitals valid, we shall have somewhat about £4,000 to divide among 8 people; had the will been null we should have seen upwards of £1,000 each.* This is all I know at present: and I thought you would be glad to hear what is doing. I have sent to John Pott's surgery a fine quarto book. Your townsman Mr. Livesey is now under Mr. Charles White at Manchester; he writes to brother Ben. and gives a very good account of your son;—"that omnes omnia bona dicere, et laudare fortunas tuas, qui filium haberes tali ingenio præditum."

I have seen a copy of Mr. Pennant's new edition of 'British Zoology,' but he has put the matter into some

* Whether Gilbert White and his brothers and sisters ever received any of this money does not appear.

stranger's hands, and has left standing many old errors; so that many sheets must be cancelled; and I must correct over again what I have corrected 'til I am quite sick!! The printing also is very incorrect.

Pray write very soon to *this place.* Respects to my sister. Nanny White improves every day. Brother Ben. has a good house at South Lambeth, elegantly furnished; but the water at present is deep in the cellar. Y[r] loving Bro.

GIL. WHITE.

On this letter Professor Newton sends the following note:—

"The 'new edition' of Pennant's 'British Zoology' referred to above is that which on the title-page is called the *fourth,* and it appeared in 1776 in two forms—octavo and quarto—with the same date, and each has on the title-page of its four volumes, 'Warrington: printed by William Eyres, for Benjamin White.' It is pretty evident that the printer must have set up the type from a copy of the old text, in which had not been marked the numerous additions to and corrections of the same, which Gilbert White had communicated to the author in the course of the letters which now form the bulk of the first part of 'The Natural History of Selborne.' The fact of White's having corrected the proofs of this edition is sufficient explanation of the omission of his own name in passages where it would have been expected to occur; for we may be sure he would be the last man to intrude it. It was thus left for Pennant to acknowledge the assistance he received from White among others in the Preface of the work; and the enormous improvements due to White's co-operation can only be appreciated by those who compare this fourth edition with its predecessors, and at the same time bear in mind the contents of White's letters to Pennant as subsequently published."

It seems a pity that Gilbert White's really great services to Pennant should only be recognised by the inclusion of his name in a list of twenty-four "learned and ingenious friends," to whose "kind informations" Pennant attributed "every merit that may appear in the present edition, and every error that may have been suppressed from the former."

To the Rev. John White. Thames street, March 5, 1776.

Dear Brother,—Thanks for your long and communicative letter. You will be pleased if you approve of the step we are taking towards the business of administration, to sign, seal, and deliver the enclosed letter of attorney, and see that it is witnessed by two persons: Brother Tho: and Ben: as stationary parties, will be so kind as to act for the whole brotherhood. Be pleased to remember that tho' our *last* council strongly asserts that we have a full claim to all that was intended for the seven lying in hospitals; yet be pleased to bear in mind also that our *first* council said also that the whole will was valid: so that should the Lord Chancellor decree away the whole, we must not be chagrined and disappointed. Three or four hundred pounds apiece are worth trying for, and will dedamage Harry from his building; and do many other good feats. Return the instrument when executed. Brother Thomas and I both think that you should *yourself* write to the Archbishop [Cornwallis] one of your best letters, and beg to know of him whether you might dedicate to him, and tell him the reason why; and then you will act on sure grounds. If you are not permitted you might mention the General in your preface.

Brother Henry went away in January before the snow was melted, and went through between high walls of that meteor. Brother Thomas by no means approves of your title of 'Zoological Anecdotes'; he thinks the latter too mean and unworthy a great book. He rather thinks that you should say, 'The Natural History of the *quadrupeds, birds, fishes* and *insects* of *Southern Spain*,' with &c. We wish also that you would throw something *savoury* into your title-page concerning *migration*; for many readers pay attention to that circumstance, without regarding any other parts of Nat. Hist. Say what you can concerning *vegetation*; for the love of such knowledge increases. Even Bishops (your Bishop in particular) in order to recommend themselves, study botany. Mr. Curtis says, that men from the other end of the town call on him in their coaches to desire private lectures for grown gentlemen. But your bookseller, at last, will be your *best* adviser respecting a title page; for such men best understand the pulse of the publick.

Pray write to Selborne not long hence.

Jack should send me an account of the thermometer at Manchester. Mr. Lever has custom at his shop; but whether adequate to his boundless views, no one can guess. He is furnishing the *whole house* with specimens; still

"Some Demon whispers, Bubo,* have a taste."

Brown, I think, is in gaol in St. George's fields; but artists never work more steadily than when under confinement. Forster has just received a letter from Linnæus, who wants to publish a new mantissa of plants in England; Brother Benjamin declines meddling. Forster says that when you write to Linn. you should direct to him not as professor at Upsal, but as *academician*; since all such letters go free, because the academy is of royal institution.

Forster's new genera of Antarctic plants do not sell; so

* The allusion here is puzzling. Why "Bubo" should have been substituted for "Visto" does not appear. Possibly Lever may have been owl-like in appearance, or perhaps it was a misquotation from a momentary lapse of memory.

that the *Skinneria*, the *Sheffieldia*, the *Barringtonia* are like to sleep. Botanists think they shall never see the originals; and other readers care not a farthing about the matter.

I marvel at the mildness of your weather in January. I told you, I think, that on the 31st my thermometer was ½ below 0°. There seems to have been a peculiarly severe current at Selborne. Mr. Yalden was so supine as never to put his thermometer out! The thermometer at S. Lambeth was at 6°; at Fyfield 15°; at Lyndon 19°! in London areas at 20°.

We join in respects. Y[rs] affect.,

GIL. WHITE.

I return home next week. Nanny White rides out at her father's country-house every day; and improves wonderfully. Molly White is at Cambridge.

Sir Tho. G[a]teh[ou]se is ruined, and locked up in a garret here in town for fear of his creditors; and his lady, who brought a £1,000 per ann. will, for his life, be reduced to poverty! Your widow-annuity is paid.

To Thomas Barker. London, Mar. 20, 1776.

Dear Sir,—You will be pleased to remember that we always mentioned Mr. Holt's money concerns as matters of the greatest uncertainty; and were well aware from the first that they must all undergo the scrutiny of the court of chancery, in which the Lord Chancellor may decree them all away, as to him shall seem most proper. But as some body must take on their administration, my Brothers Thomas and Benjamin are willing to undergo that troublesome trust, thinking it best not to leave a business in which our family is so largely concerned to strangers, or insufficient people. The Judge in the Commons has already determined, upon the application of Mr. Tawke in which he had no concern, that neither the Foundling Hospital, nor any of the Lying-in Hospitals shall administer, but the next of kin.

All our chalk-barley was greatly damaged: we have none good but from the sands, where their harvest was earlier. The former sells for about 20*s.* and the latter for about 26*s.* Farther west, where they have nothing but chalk-downs, *all* their barley was pretty well spoiled; so that Harry complains he cannot, as yet, get any price at all for his crop.

What pigeons in a state of nature may do I cannot pretend to say; but this I know, that tame pigeons, which are pampered by high feeding, and lie perhaps under more frequent temptations from living together in crouds, are apt to forget the rules of strict chastity, and follow too often the example of people in high life. As to the smaller birds, it would be very difficult to ascertain the identity of the man and wife in different years.

Yours affect.,

GIL. WHITE.

I thank you for this hint; and shall always be glad of more.

To Samuel Barker.

[On the same sheet as the above.]

Dear Sam,—You may comfort yourself that you are not the only person that finds himself under difficulties respecting the sexuality of mosses. For Mr. Curtis is by no means satisfyed concerning the distinctions made use of; but suspects very much that there are not the same obvious distinctions in them which in the more common plants so rationally support the Linnæan system. He is a very friendly man, and always willing to communicate; and has therefore desired me to send you down the enclosed plate containing representations of the fructification, &c. of his mosses, such as he uses in his own lectures on the occasion. In the plate respecting the male and female *Vallisneria* you will see a wonderful instance of the wisdom of providence. Mrs. Stebbing* and I go down to Selborne together on Friday next in a post-chaise. I shall not forget to take my sister's present.

Yrs affect.,

GIL. WHITE.

The water-snails, two or three species, now begin to be buoyant, and to crawl with their bellies upwards on or against the surface of the water. We have none round Seleburne.

On March 26th, 1776, after his friend's return from his visit to London in January and February, where he noted in his *Naturalist's Journal* the appearance of the town "in Siberian weather," Mulso writes from Winchester:—

"I cannot tell where this will reach you; but knowing your old exactness, I imagine that you are now returned to your Curacy, as your tether has been stretched this time more than usual.

I hope you are apprized by this time of the value of your share of the assets of Mr. Holt, and have found them rather to exceed than to fall short of your expectations. It is well when so good a fortune falls into the hands of so good a man as your brother Thomas.

I have hardly heard from my brother or sister Mulso this winter. I only heard from Sister Chapone that you had been at their house. I therefore shall be curious to know whether you fixed with my brother upon any plan for your views and drawings. The lateness of the season now makes me suspect that your work will not come forth

* Mrs. Etty's sister.

this spring, and indeed the want of that ornament, which you seemed to set your heart upon, will make it impossible. I feel an impatience, and the more for your sake, as the taste of the town for reading is capricious, and natural observations have had a run, and at a high price. I should rather therefore have wished both you and John to have pushed your collections forward. But you must be best judge now you have been at London."

In March Gilbert White returned to Selborne, whence he wrote—

To Thomas White. Selborne, April 6, 1776.

Dear Brother,—Though I must not aspire so high as to make any pretensions to the rectory of Gotham; yet may I claim at least to be Mr. Lightfoot's curate: for I have acted so simply by Berriman's bill, which I brought up to be forwarded to Mr. Cane, that unless you can help me out I shall, in all appearance, be ten pounds out of pocket for my meddling. Now this bill, which was a Guildford note value £10, did I carry in my pocket intending daily to pay it, with £1. 5. 0. more, into the hands of Bro. Benj. Somewhat still, forgetfulness I suppose, prevented my paying it to him. However in some moderate time the bill was gone; and I had so fully possessed myself that I had done what I ought, that I thought little of the matter; only now and then asking my Br. Benjamin if Mr. Cane had drawn for the money; to which he always answered in the negative. Under this perswasion I came down and told Berriman that I had *payed* his money. When, lo, in a post or two comes a letter from Mr. C. complaining that he had drawn for the money; but that the answer was that no such money had been paid in: from whence I conclude that my Bro. is not conscious that I ever paid him: neither indeed can I *say* that I *did*; only that I *thought for weeks* before I left him that I *had.*

Now as you see the state of the case, and how simply I have acted, pray recollect all you can concerning a £10 Guildford bill, drawn *for*, and backed *by* Robert Godard, N° forgot. Did I ever change such a bill with you? or did I ever talk before you concerning any disposal of said bill. To the best of my knowledge I still took cash of you for *all* my occasions, and payed for *all* my necessaries in cash. Could I leave it behind in my drawers, or in any book? All my other concerns come right except this unfortunate bill: so pray help me out.

The old attorney at Petersfield told me that *all* the Holts were dead, or gone from the town before his memory of people: but a Mr. Holt, he says, *who was agent to the Duke of Bedford,* came to him in 1729, and desired him to dispose of a farm for him, which lay in the parish of Hawkley. This farm, the Gent. told me, he did dispose of for Mr. Holt. Moreover, he told me that the estate called Nursted in the neighbourhood of Petersfield, value about £300 per ann. was for generations in the possession of the *Holts*; and that he believes the Holts of Petersfield sprang from the Holts of *Nursted*: that there is now in a window in Petersfield the *arms* of the Holts in painted glass brought from Nursted: that the last Holt of Nursted was member for Petersfield many years in the reign of King William: that he had only a daughter, who marryed into Dorset: that afterwards the estate, Nursted, was sold to Mr. Hugonen the Swiss, whose son possesses it still. Mr. Etty has paid *me* your bill: but he expected to have been charged for a parcel of yellow ocre, which was not in said bill. The farm which I sold at Harting, Scot's,* is now on sale again. I sold for £1,120: and they ask £1,600: but they have built a *new* house on it. We are going to build

* It was sold in 1761. No doubt this was one of the properties realised to pay the younger children's portions after the death of John White (Gilbert's father) in 1758.

a new Hermitage. You have sweet weather for your Journey. I must go to Oxford on Monday: but must return on Saturday. My evergreens have suffered very little: Portugal laurels not at all.

Yrs affect.,

GIL. WHITE.

Please to pay the Master of the post chaise in the Borough for my fare to Hounslow: the boy knew nothing of the price, and the people there told him 10 or 12 shillings.

I have got the register-account from Rogate of our Father's and Mother's marriage, such as it is: but it is very blind.

To the Rev. John White.

Fyfield, May 15, 1776.

Dear Brother,—Brother Thomas and his daughter have been with me for a week: yesterday we all three came down to this place, and found all well.

As I had got a frank for you, I thought it best to take it down with us, that all that had occasion might make use of it together.

Forster was presented to his Doctor's degree at Oxford on account of his literary fame; and because he had hazarded his life in a circumnavigation in the pursuit of Natural knowledge.

Pray write to Linn.; for if he only tells you of your new genera, &c. and affixes no names, he leaves you in the lurch; direct to him as academician for reasons assigned in a former letter.

We are glad you find your heavy duty so easy; for what is paid to curates is all neat money, and occasions considerable deductions from a moderate living.

Holdsworth I have procured; but I can't say the work gives me so much pleasure as it seems to have afforded you. I did not find so many genuine criticisms drawn for the face of the country, and the modern practices in husbandry, as I expected; but rather a collection of parallel passages from Cato and Columella. So much easier is it to compile, than to advance fresh remarks.

Mr. Yalden will not probably set out for Spain 'til next spring: I shall exhort him by all means to take your recommendations in his pocket. He is a decent zoologist; and particularly an entomologist. He is returned, I hear, for Edinburgh to Winton.

Your quantity of drawings, I find, are considerable; no doubt they should be engraved in London.

As to your botany, it should be carefully overlooked by some body: in the zoological part your powers are much more considerable; and you want only a friend, as all men do, just to remark those small errors, or slips which *incuria fudit*.

You are the best judge whether you should address Mrs. Cornwallis; if you do, you may express your regard for her husband with more warmth than if he was living.

"Quis *desiderio* sit pudor aut modus"?

Send me some account of the *Hippobosca hirundinis* at your leisure.

Harry's little academy is in a flourishing way: he will, it is too probable, lose his £100 young Gent. at Midsummer, but then at that time he is to have a fresh pupil of 14 years of age at £150 per annum!

His building has been heavy on him; but without considerable additions he could not have stood on the present footing. So one must be set against the other. The common parlor at present is the worst story; indeed it is most sadly crouded: however brother Thomas is going to make an addition to it of a space of 18 by 10 feet, which is to be built on towards the street, with a window looking down the street, and the chimney at the upper end: when finished the parlor will be thus—

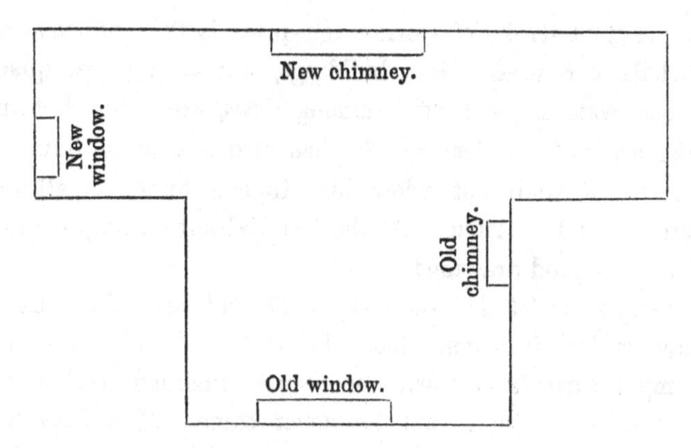

like a T; odd, but roomy and convenient.

We are much distressed in Hants by the long dry season: no grass, and a poor prospect for spring-corn: butter 10*d.* per pound, and hay at £4 10*s.* per ton!

Mr. Chandler's stile and wording are very lame and defective indeed.* Sir Thomas Gatehouse's effects are just sold off at both his seats.

My respects to my sister, and Jack—John I mean, now he is six feet high.

You cannot take an other living without becoming A.M. or LL.B., both which degrees will require time, attendance, and expence; if you take a second living now you render your first void *ipso facto.*

The bloom on all sorts of trees is this year very extraordinary indeed!

At present I think of sending for Grimm about the beginning of July: I may employ him for perhaps a month. Mr. Yalden of Newton then talks of taking him for a week to draw his house, and outlet; and then he is to go to Penruddock Wyndham Esq[re] at Warnford. So he will

* This may possibly refer to 'Travels in Asia Minor,' published in the previous year (1775), or to 'Travels in Greece,' published in 1776, by Richard Chandler, the traveller and antiquary, with whom Gilbert White subsequently became intimately acquainted.

have a good stroke of work. His price is two guineas and an half per week. His buildings, human figures, quadrupeds, waters, perspective among trees, are good; but his trees are not so pleasing: he has also a vein of humour, but that I shall not allow him to call forth, as all my plates must be serious. At the last Exhibition he produced some very good drawings.

Harry's outlet is now very neat and beautiful. Capt. Shaw is but just gone from Farnham. I called on him in my return from town; he seemed inclined, I thought, to make me a visit; but he never came. Mrs. Etty has been very ill indeed since her lying-in, but is getting better; she has got an other son, whose name is Simeon.

Writing on May 5th, 1776, Mulso remarks:—

"I shall envy you prodigiously when Mr. Grimm is with you. What a plenitude of Virtu will you feel within you, re-creating Selborne and immortalizing your Favourite Place."

On May 27th, 1776, Thomas White heard from his brother, who was now busy with 'The Antiquities of Selborne.'

"Pray desire brother Benjamin to enquire of Mr. Astle and others where they suppose I can find Chart. 54 Hen. 3, m. 3, *De mercatu et feriâ apud Seleburn*, whether in the Tower or elsewhere.

"We have finished the hermitage, and I desire to know what I am to charge you towards it, as you were so kind as to make a tender of your subscription."

This building, called by Gilbert White a "new Hermitage" in a recent letter to his brother, may be seen in Grimm's large north-east view of Selborne, in the middle of the picture, which formed the frontispiece to the first edition of 'The Natural History and Antiquities of Selborne'; the original Hermitage,

> "... the scene that late with rapture rang,
> Where Delphy danced, and gentle Anna sang,"

appearing more to the left, higher up, and near the zigzag path up the Hanger.

To Thomas White. Selborne, June 24, 1776.

Dear Brother,—Molly says in her letter to Mrs. Etty that you think of going into Rutland next month. I am going to write to Grimm, and shall desire him to come down about the 8th of July. We have a black cold and blowing solstice; but not a wet one. My S^{t}foin was cut last Thursday, and will be ricked tomorrow in good order, if the weather holds out. If you don't write soon and forbid me, I shall set you down as a subscriber to the Hermitage, which is finished.

Mr. Butcher* is a very extraordinary man: but you seem a favourite now. He puts me in mind of Sarah Duchess of Marlborough, whose resentment Mr. Pope says was the most formidable thing in the world—except her favour.

Harry and his wife will I apprehend find much trouble with the measles; but the young folks can never have them at a better time.

I am glad the Lady in Herts continues so gracious.

Pray remember my *records* at the Charter-house Westminster. I have just ceiled and am fitting up a garret for any young person I may have with me: that there may be more room for Molly and yourself when you come. . . .

Love to Molly. Mrs. Etty improves daily.

Yours &c.,

GIL. WHITE.

* The family attorney in London.

To Samuel Barker.

Selborne, July 1, 1776.

Dear Sir,—In the larger plants in general I found no difficulty at all, being assisted by Linn., Hudson, Ray's 'Synopsis stirpium,' &c.; but as to the mosses, I did not care to meddle, because they were become too minute for my eyes, before they began to be employed in those enquiries. As to the genera of *Orchis*, *Ophrys* and *Serapias*, there is great obscurity among them, as Linn. tacitly acknowledges by calling in the distinction of their *roots* in his specific descriptions. In his system the distinctions should lie in the corolla, stalks, leaves. As to the different sorts of garden-fruits, they are the production of cultivation; and belong only incidentally to the Linn. system; but are to be sought for in 'Miller's Dictionary' &c.

As to the *Nymphœa alba*, the white water-lily, there are none in this parish: but are to be found in Wish-hanger pond in the parish of Frinsham: if I knew when the seeds were ripe, I would endeavour to procure some.

Dr. Forster assured me that he saw, some few times, appearances in the sky in the S. hemisphere very similar to those we call *auroræ boreales*.

Our *first* young brood of swallows was marvellously early, appearing on June 15th: the first week in July is the usual time. I have written to Grimm, and don't know but that I may see him next Monday. I expect also my nephew Richard White, who is to stay with me some time. Nanny Woods dines with me to-day: she is tender, and has got a cough.

We have built a new Hermitage, a plain cot: but it has none of the fancy, and rude ornament that recommended the former to people of taste: this is strong and substantial, and will stand a long while, fire excepted. Our solstice is cool, and shady, but not very wet: I have ricked my S^{t}foin in fine order: this day I begin to cut my meadow-grass, which will prove a bigger crop by one third than that of last year. Cucumbers do not succeed well in general this year. My bank, which was lowered when you were here, is now very gaudy, and full of flowers. I have much wall-fruit; and a fine show for grapes: pears, plums, apples, and cherries without number. As I was visiting last Tuesday at Bramshot I saw on the Portsmouth road Burgoyne's light horse marching down to embark for N. America: the horses were fine, and the men fine young fellows; but they all looked very grave, and did not seem much to admire their destination. The Atlantic is no small frith for cavalry to be transported over. The expence will be enormous! Brother and Sister Harry have been in town, and at Mrs. Snooke's. At Ringmer Ben. Woods caught the measles of John Mott, and fell with it before he left town: but his father sent him down to Fyfield after it came out. None of brother Harry's children nor brother Thomas's have had this distemper: so there will be a sick house, and much trouble: but the children can never fall at a better season of the year, or time of life. Now at midsummer Harry is to have a young gentleman at the noble price of £150 per ann.* Harry and his wife (no small personages) and seven children, two canary birds, one aberdavine, 10 parcels, a dormouse, and a puppy-dog, all went down in two post-chaises. Brother Thomas is going to enlarge Harry's common parlor by building on a large piece at the N.W. end next the street: the room then will be odd but large and convenient: it now swarms with people. H. Woods and Bess† make some progress on

* A Mr. Amyand, apparently. An account of Henry White's family, with extracts from his diaries containing much interesting information upon country life in the eighteenth century, may be found in 'Notes on the Parishes of Fyfield [etc.],' by the Rev. Robert Hawley Clutterbuck, F.S.A. Bennett Brothers, Salisbury, 1898.

† Elizabeth, daughter of Henry White.

the harpsichord. Mrs. Etty was very ill in the spring; but is recovered. Mr. and Mrs. Yalden are at Sarum attending on the nuptials of Molly Fort, the youngest lady.

Y^rs^ affect.,

Write soon. GIL. WHITE.

On July 8th, 1776, the *Naturalist's Journal* records—

"Mr. Grimm, my artist, came from London to take some of our finest views."

Writing on July 16th, 1776, Mulso regrets his inability to visit at Selborne.

"While you are enjoying yourself, like an Italian magnifico, with your designer at your elbow, I am waiting for an artist in his way, that may be perhaps as profitable, but is not half so agreeable to my taste—I mean a surveyor. . . . You may imagine that you whet my curiosity by telling me what a pleasure I might partake from Grimm's pencil, and the liberality of his manner: and it would be no small part of my satisfaction to see my old friend taking such voluptuous rides upon his hobby-horse. No man communicates the pleasure of his excursions, or makes the world partake of them in a more useful manner, than you do. It is the

'Solemne viris opus, utile famæ,
Vitæque et membris.'

"Your work, upon the whole, will immortalize your place of abode as well as yourself; it will convert men's principles; and give health to those who chuse to visit the scenes of Mr. Grimm's pencil, in their original. I have a good opinion of the correctness of Mr. Grimm's fancy, by what he judges of my brother Mulso. . . . I have seen something of what you mean of Mr. Grimm's

Tinges * in some little things of Taylor and others at the exhibitions in Town; it is exceedingly pleasing, and could you have it transcribed into your prints, would wonderfully improve the force of the drawing. I long to be at your side, but I cannot: yet by the time that the first insanity is over, and you begin to speak slower and in a milder voice, I hope to talk with you here or at Selbourne.

"Do you consider that I am now ploughing? that hay-harvest is coming on? . . . Add to this visiting and visitationing, swearing at Quarter Sessions, and all the wickedness and dissipation of Plurality? †

"Well, my dear friend, I have much to be thankful for. . . . . I am glad to say I left the Bishop well enough to seem likely to serve several friends more. I *could* direct him—but it is not allowed. Have you read Gibbon's book? What think you of his latter chapters? If you dislike them, cannot you answer them? You have the candour of a gentleman, and could confute a genteel writer in a decent way. I *wish* you could; and *soon*: you have leisure, and you have access to what books you might want. (Among ourselves—*in illumquidem beneficium collocarem, a quo graviter lucide et viriliter convinceretur, i.e.* G. This was the word, if I express it right, of him who seldom breaks it.)"

Even this bribe, however, did not move the historian of Selborne to enter the lists against the historian of Rome.

At the beginning of August his friend visited Mulso at Meonstoke, whence he wrote—

* The tinged drawings subsequently spoken of by Gilbert White in his letter of August 9th, 1776, to John White.

† He had recently been presented to the Rectory of Easton, near Winchester, by the Bishop of Winchester.

To the Rev. John White.

Meonstoke, Aug. 9, 1776.

Dear Brother,—By your unusual silence I began to fear what has really been the case, ill health. You have perhaps by your attention to your book and other matters been too free with your constitution lately: you must therefore relax a little, and allow yourself more time for riding and walking. Particularly, I think, you should avoid contention though in ever so good a cause: for any earnest agitation of the mind is bad for the stomach and bowels.

Luccomb's oak, we think, will probably turn out at last the *Quercus ægilops*: but this matter cannot well be determined 'til it comes to bear fruit. It carries its leaf all the winter in Devon, but casts it at Selborne, Essex, and elsewhere; and is probably a deciduous tree.

Perhaps your *homo sapiens* may be too close a copying of the Linn. system, and may appear pedantic to an anti-Linn. reader. I by no means want the *Hippobosca hirund:* just at any one particular day or week; only wish to see it at your leisure. Mr. Grimm was with me just 28 days; 24 of which he worked very hard, and shewed good specimens of his genius, assiduity, and modest behaviour, much to my satisfaction. He finished for me 12 views. He first of all sketches his scapes with a lead-pencil; then he *pens* them all over, as he calls it, with indian-ink, rubbing out the superfluous pencil-strokes; then he gives a charming shading with a brush dipped in indian-ink; and last he throws a light tinge of water-colours over the whole. The scapes, many of them at least, looked so lovely in their indian-ink shading, that it was with difficulty the artist could prevail on me to permit him to tinge them; as I feared those colours might puzzle the engravers: but he assured me to the contrary. From me Mr. G. went to Mr. Yalden to take a scape of his outlet from above the chalk-pit. On Tuesday I brought my artist (at the desire of a gentleman who was visiting there) to Lord Clanricarde's at Warnford, that he might take a drawing of an old building in his lordship's garden, now a barn; it is a curious piece of antiquity little known, and will prove an agreeable surprize to many as I am sure it was to me, who never heard the least of the matter before.

Brother Thomas can by no means yet bring his matters to a conclusion; he is lowering his stock, and preparing to retire; and will hereafter probably spend part of his time at Fyfield, as he is building there. He has just carried his daughter Molly down to Lyndon. Brother Henry has just had the measles in his house: twelve young people went through that distemper, and all are well recovered: his £150 pupil was to have been with him at midsummer, but has been kept away by the above-mentioned disorder. Mr. Halliday is with him yet, but may probably go at Xmass. There is some reason to fear that that young gentleman's father may be taken by the American privateers in his passage home from Antigua.

Mulso has just got a second living near Winton; the name of it is Easton, it is worth £250 per annum. Our S^t^foin was finely made; then we had a dripping time that spoiled much clover, and some meadow-hay; and for the last fortnight in July we had glorious weather to finish off the meadow: now harvest is beginning, and the weather dripping. Mr. and Mrs. Mulso join in respects. I saw one swift yesterday. At present I cannot say when I shall be at liberty to wait on you and my Sister; but you may be assured that I wish to have it in my power to see Blackburn.

I conclude y^r^ affectionate Brother,

GIL. WHITE.

A little later on the usual visit to Ringmer was paid.

To Samuel Barker. Ringmer, Aug. 19, 1776.

Dear Sam,—A knowledge of the grasses is the most desireable part of botany, because the most useful; but it is the most neglected, for graziers and farmers do not seem to distinguish any one sort of *Gramen* from the other; the annual from the perennial, the succulent from the dry, or the aquatic from the upland. Whereas by attention their meadows and pastures might be much improved; and it is an old maxim, that he is an useful member of society, a good common-wealths-man, "who can procure two blades of grass where only one grew before."

I am not a little pleased to find that you have got the *Hirundo riparia* just at hand; because I shall expect from a man of your accuracy some circumstances of information, which I cannot so well make myself master of at the distance of Wolmer forest. You will be pleased, another year, to attend to the *exact* season of their coming and departure; time of nidification, and bringing out their first and second brood, &c., &c. Pay attention also to the other three species, for I shall be glad of *any well-attested* anecdotes, intending some time hence to publish a new edition of my '*Hirundines*' in some way or other. My artist Mr. Grimm stayed with me 27 days; 24 of which he worked very hard, and displayed great tokens of genius and assiduity. He finished twelve drawings; one of which was a view of Hawkley-hanger; but that scape, I think, did not succeed so well as some others. Bro. Harry made me a visit, and was much delighted with what was going on; and in particular with a view of the hermitage, which Grimm is to copy for him in town.* From me my artist

* This was done. This really beautiful water-colour drawing of the original Hermitage, which is considerably larger than the vignette published in 'The Natural History of Selborne,' is now in the possession of the Earl of Stamford, a great-grandson of Henry White.

THE HERMITAGE AT SELBORNE

FROM A WATER-COLOUR DRAWING BY S. H. GRIMM, 1777

[To face p. 328, Vol. I.

went to Mr. Yalden; and took a view of his house and outlet from the edge of his chalk-pit. The employer wanted and intended a view from the alcove; but the draughtsman as well as myself, objected much to the uniformity of that scene; so I carried G. to the chalk-pit, on the W. side of the house, from whence he took a charming view. From Newton I carryed G. to Ld Clanricarde's at Warnford; where in the gardens he took a perspective internal view, section, and elevation of a very curious old hall, or church unknown to the antiquaries, for a gentleman visiting there, who will one day oblige the world with this neglected and obscure curiosity, now a barn. It is supposed to have been built by King John: the order is Saxon. From hence G. went to Winton, to work there for a week or ten days on his own account; and is to call at Harteley on his return. I regret much that King John's hall had not remained unnoticed a little longer, 'till my brother Thomas had been a little more at liberty; for when he has done with business, he proposes to entertain himself with collecting materials for an history of the county of Southampton; and I moreover marvel that I never heard of this hall before.

Our people know nothing of the use of the rind or peel of the *Juncus effusus* for cordage. I rejoice to hear that you learn French; you will very soon be able to read it.

The weather has been now very showery for just a fortnight: our harvest is in a very bad way. When I arrived last Friday evening I was surprized to find Mrs. K. Isaac and Niece Becky at supper with my aunt. B. is grown beyond all knowledge. Nephew Richard, who has left school, is here with me.

Mr. Shadwel has left Stoneham farm, which is raised from £250 to £400 per ann. Thanks to Molly White for her agreeable letter. On the Friday you mention my thermometer rose up to 79° *within* doors; a pitch which

I have scarce ever seen exceeded. Wheat grows in the sheaf.

You have the *Stoparolæ*,* I find; but say nothing of the white-throat, black-cap; *Reguli non cristati*, 3 species; the red start. Respects to all.

Yours affect.,

GIL. WHITE.

We have just weighed Timothy, who is increased in weight just one ounce and an half since last August. *Stoparolæ* come to Selborne May 20th and depart about Septr. 7th.

* Now known as the spotted or grey fly-catcher.—A. N.

END OF VOL. I.

WILLIAM
BRENDON & SON
PRINTERS PLYMOUTH

Printed in the United States
By Bookmasters